Siesta
Lane

Endangered Species*

Sometimes I feel a diminishing
as if part of me were being erased
and I wonder what in human nature
will disappear the moment copper
no longer singes the coat
of the grizzly wading old growth
when the pelican never again
trowels the clouds
above the spray and barnacled back
of the blue whale surfacing into sky
like a ferruginous hawk rising
from the sinews of its prey
spiralled amid the crystal quiver
of prairie grass
where a silverspot butterfly
places its final tear of bubbling larva
while the last salmon flashes upstream
jostled by jets of sperm
past the cut-throat trout
sashaying in rock shadow
beyond where the mariposa lily
folds its petals forever
as the lone western pond turtle
drags its humped knees
to the frog-loud, lotus-speckled
oxbow pond that has not yet
been filled for a parking lot
where you will leave your car
hurry to your office sliced by anxiety
that you have lost something,
but you are not sure what.

*All species mentioned are currently threatened.

Siesta Lane

One Cabin, No Running Water, and a Year Living Green

amy minato
illustrations by jan muir

Skyhorse Publishing

The places, people, groups, and events in this story are real,
although filtered through the author's gymnastic imagination.
Names have been changed to protect the shy.

Skyhorse Publishing books may be purchased in bulk at special
discounts for sales promotion, corporate gifts, fund-raising, or
educational purposes. Special editions can also be created to
specifications. For details, contact the Special Sales Department,
Skyhorse Publishing, 555 Eighth Avenue, Suite 903, New York,
NY 10018 or info@skyhorsepublishing.com.

www.skyhorsepublishing.com

10 9 8 7 6 5 4 3 2 1

Library of Congress Cataloging-in-Publication Data

Minato, Amy.
    Siesta lane: one cabin, no running water, and a year living
    green/Amy Minato; illustrations by Jan Muir.
        p. cm.
    ISBN 978-1-60239-328-8 (alk. paper)
    1. Communal living—Oregon—Case studies. 2. Country
    life—Oregon—Case studies. 3. Minato, Amy. I. Title.
    HX655.O7M56 2009
    307.77'4092—dc22
    [B]
                                    2008029537

Designed by LeAnna Weller Smith

Printed in the United States of America

To my dearest
Joe, Mateo,
and
Ruby Lou

Northwestward Bound

It came to me during a Chicago traffic jam, trapped in a metal oven, a popsicle gooeyed onto my sweaty thigh, the guy in the next car cussing through his cigar. I gazed through steaming sunglasses at this muscular city buried in concrete, its veins plugged with cars, and felt my life constrict like a capillary

Now, I love this city with its fashion parade of architecture, the gruff but generous people, Lake Michigan with her necklace of public beaches. Yet every spring migrating birds confuse reflections of clouds and trees with real nature and collide against skyscraper glass. I imagined flocks of warblers, sparrows, thrushes struggling through the pollution above my windshield, disoriented, smashing against panes—stunned or dead. In my mind's eye I was a migrating bird—swerving across the freeway's metallic seascape amid plumes of smog trying to discern truth from illusion.

There were warning signs even then about greenhouse gases and climate change. Higher temperatures shrink ice caps, increasing water surfaces that absorb more heat, producing higher temperatures that raise sea levels in an accelerating cycle. I thought about all the people living on coastlines, holding their breath. About the planet poised in this crisis like a glass of water about to tip. While surrounding me at just this one moment on just one day in just one of hundreds of cities on earth, thousands of vehicles were spewing toxins into the air like grey cotton candy.

My fellow drivers began a crescendo of honking not unlike a cricket chorus or the alarms sounds of small birds when a hawk circles like a call to action. What is the ethical response to knowing that my lifestyle threatens the health of our planet and our species' survival? Could living simpler make a difference to the future? To my sanity? In an age of dwindling resources, maybe we'll all have to downsize anyway, so it could be empowering to try it voluntarily.

At that moment I began a journey against the consumptive current—to seek my own refuge from which to ponder this dilemma, to heal some of my longing for wildness. I would find a place where cougars roam and a few clean rivers still carve their own channels.

Wasn't Oregon at the heart of the sustainability movement? Organic food. Green culture. Big trees. Mountains. Loud honks and "Whatya waitin' for, lady? The Cubs to win the pennant?" broke my reverie, but I began, in my mind, packing to go West.

Siesta
Lane

Sojourn

Two years later I'm 30,000 feet up, cradling a copy of *Walden*, on my way to graduate school in creative writing in Eugene, Oregon. The aircraft I'm riding in takes its form from nature, but the amount of fuel it uses makes flying one of the least sustainable modes of travel to begin my more sustainable life. It's my first step on a tremulous tightrope toward my goal, from which I will slip, regain balance, slip again. Not so simple—simple living. I console myself by figuring that, if all works out, I won't need a return ticket.

Clouds mingle like party guests, like my own fears, as we cross the Great Plains. The land below has been denuded, sliced into jigsaw pieces of various shades of brown. I search for the topsoil—the moist, rich, black soil on which most terrestrial life depends. What I see is a barren landscape. The occasional rivers pinch the land like metal strips. A few yellow trees line their banks. Why are they yellow, I wonder. Cottonwoods growing near water? Or poison from a city upstream?

But it may just be that the inner West is a thirsty place. Too often pioneers that settled a claim in a fluke wet year were scraping at dry soil when the regular climate swung back at them like a scythe. Would I suffer a similar fate? What realities might dessicate my dream? Each tilt of the plane swung a pendulum between my pragmatic and quixotic natures. Finally, the mountains come into view—large, resting mammals. Bald patches from clear-cuts scar their sides.

The stewardess reminds us that the movie is starting. *City Slickers*, about three men who leave New York to vacation at a dude ranch. She tells us to close our window shades to better see the picture. When we imagine the American West, it will be a movie of the frontier we see, film taken of the same virgin area again and again. Close-ups of a few sagebrush. Buffalo brought in for effect.

Not the real story out the window.

We land in mist-wrapped Seattle and I board a train to Eugene that stitches through the Willamette Valley—the end of the Oregon Trail—with its volcano-enriched soil, soggy logging towns, and the roil and murmur of grey-blue rivers. Dogwood blossoms peer in. Douglas firs press green cheeks against the glass. I rock along with the wheels winching on the tracks, catching glimpses of strangers across the red seats. Who are they? Are any of them traveling into a new life? Are they gathering their courage as well?

The Search

Neighbors stare at me blaring blues from the porch of my garagetop studio apartment. Homesick one night, in this new age Mecca with its hippie roots, I sip whisky, fiddle with a cigar and choke down steak in a gesture of loyalty to Chicago, never mind that I did none of those things there. It helps with the lonelies.

Soon enough I am a yoga-practicing, tai chi, chai tea vegetarian eating tofu on my futon. I finish grad school, dive into a medley of work and social causes, travel, and swear allegiance to a motley crew of buddies. But though a hip college town, Eugene is still a city where buildings obscure land. My soul pinches and pinches at me until I search the classifieds for a way to live more rural and find two possibilities.

"Rustic cabin on creek near wildlife refuge," says the first rental ad. I bribe my New York professor writer friends, Sophia and Gabriel, recent University of Oregon recruits, who claim my siblings pay them to look out for me, to come along and provide levity. While, after eight years, I find Eugene too urban, they struggle with it as a backwater and agree to join me because the idea of living more rural than 100,000 people both intrigues and alarms them." Won't you be freaked?" "What about restaurants?" "There aren't even taxis!" After a long drive on a tire-eating road, we arrive at a farmhouse with shells of trucks in the yard. A large woman holding a bowl the size of a tire dumps its mountain of food scraps into a chicken coop on her way from a rickety porch to our car "Hey, there!" She wipes a hand on overalls mottled with dog and extends it to me. "I'm Pat. Come in for some coffee."

"Actually, we just have time to see the cabin, if that's okay, but thanks," Gabriel replies, gawking at a large bull in the unfenced yard

Pat shakes her head, sighs, but smiles at us, "City folk, always in a rush."

We walk a narrow trail through dense vegetation to what can only be called a shack with broken windows. "Rustic," in this situation, translates to "fire hazard."

"It's real cool in summer," Pat offers.

"I bet," Sophia agrees, with a sudden desire, I imagine, to be stirring an Italian soda with a tiny straw in an upscale Manhattan café.

Inside the place is bare and dusty but "not without potential," I convince myself, thinking a good sweep could do wonders. "Let's see the creek," I declare. Pat leads the way.

We hack our way to a tiny ditch where newly hatched mosquitoes swirl above the few wet spots.

"This is high water season. It'll dry up in

summer, when the rain stops." Pat says, her boots deep in muck.

Sophia untangles her sweater from a blackberry bush and Gabriel begins walking back.

I, however, am not yet willing to give up my dream of living closer to nature. I know there will be challenges.

"How did the windows get broken?"

"Well," Pat chuckles, "I figure the bear did it."

"The *bear*." Gabriel's dark brows dance a tango.

"Yeah. The last renter used to feed this bear right from the cabin. So now the bear thinks she has rights to it. She breaks in every now and then."

We reach the shack, "You can see the claw marks here on the wall."

Sophia bolts and Gabriel pulls me by my jacket sleeve. They both call: "Goodbye Pat, thank you."

"No need to be scared," Pat assures us. "Bears don't bother with people; they go after your food. So are you interested in the cabin?"

"I'm looking at several places today. I'll let you know," I say hopefully.

"Amy, you're not living there," Sophia states when we get back to the car. This coming from Sophia, the world's most gracious and lovely woman, gives me pause.

"It could be interesting," I argue tentatively.

"Living under a bridge on the Jersey turnpike could be interesting," Gabriel adds, gripping the steering wheel as we negotiate ruts the size of Idaho.

In my mind, fear and intrigue battle. I like black bears, hypothetically, their sharp intelligence and long-term memory, their human-sounding voices. I'm especially intrigued by their ability to sleep for months, and I know that they virtually never harm humans, even though we are the main cause of death for them. How fascinating it might be to study a black bear up close! But the reality of one so bold near where I would be living raises the hair on my skin. Maybe wild nature is too alien for this Midwest gal. Maybe I fall short of what my dream requires?

Sophia and Gabriel cheer me with tales about the munching deer outside their first place in Eugene that frightened them into a twelfth floor apartment. Gabriel claims that a crazed squirrel terrorized him into keeping the shades drawn and Sophia confesses that animal noises at her last writing retreat in the woods chased her home early. I confess to being scared of rats. This leads to a litany of horrific animal anecdotes: snakes coming up through toilets, raccoons taking over woodsheds, leeches sneaking inside swimsuits.

Finally, we reach the paved road and have exhausted our animal attack stories. I promise not to move into the bear cabin and everyone relaxes. Frost pulls her skirts from the fields and birds hail us from fence posts. The sky is a splendid blue.

Rainbow's End

After winding past posh homes and horse pastures through aptly named Rainbow Valley, we follow a long driveway off Siesta Lane to the second potential rental nestled on a hill among oaks. A "Come On In" sign welcomes us into a tall, bright house where large windows look out across a meadow toward the Coast Range. The woodstove warms the bright kitchen. Big, colorful pillows edge the living room. Three people in Chilean sweaters are eating egg frittatas and reading last Sunday's *New York Times* around a large wooden table. Daffodils in a vase open their dog faces and give off puffs of yellow scent.

We chat awhile with the brunch-eaters, Mick, Luke, and Raul, who share their delicious coffee with us, ask our stories, and tell us about themselves. "What's the deal with this place?" Gabriel asks, cutting the egg frittata into triangles on his plate. "Who pays for what?"

"It's terrifically economical. We split electricity, phone, and water bills eight ways, share basic equipment, use wood heat in timber country, and even grow some of our own food. But it's more work, too." Mick suggests, pouring more coffee into a chipped but handmade purple mug with leaf imprints. "We write messages, problems, and solutions in a big journal called the Red Book. Then we spin a circular list of household jobs called the Wheel of Fortune to see what chores each person is assigned to do for that week."

Raul shifts in his chair. "If you mess up on your chores," he raps his plate with a knife, "Mick whips you with native beargrass. That reminds me ... " and Raul's off, clutching a broom with Luke following, pushing up his thick glasses, slipping on boots, still reading the Art Section.

Gabriel looks at me and shakes his head. "Sounds right up your alley, kiddo, and in line with your finances." Then he settles in to read the Book Review; Sophia browses the library. A goldfinch bobs through the air and plops on an oak branch beside the window.

I will later learn the basic history of the place: Sixteen years earlier Rainbow Valley, with its clay-dense soil and hard water, was home to poor farmers, trashed buildings, and rusty trucks. For $22,000, Quin Wilson bought these eight acres and built this house and a couple of the cabins for himself and his wife. Then his marriage fell apart and he rented the cabins out.

Using salvaged wood from his construction company, he built one cabin or yurt after another and rented them to friends, imagining a new kind of community. People came to the main house to cook and use the bathroom. That ensured social interaction. But the cabins, spaced apart with trees between, were built for privacy.

At one point there were seventeen folks here, mostly kids with single parents and eccentrics who finally found a sense of family. The original idea was that folks would pay rent toward owning the place and it would become a true cooperative, but that didn't work out. People would couple up and move off, wanting to nest. Quin lived here for thirteen years until that happened to him. Also, his children had become busy high-school athletes and the commute to town became tedious.

Amy Minato

A towering, quick-grinning guy with buoyant blue eyes and strong, weathered hands, Quin, Siesta Lane's Wizard of Oz, still appears at whim on the property, magically fixing a broken shower, adjusting a crooked drape, dropping off salvaged firewood.

Stained glass windows, wood carvings, and other artistic touches express Quin's aesthetic vision. The main house pantry with four refrigerators, a bathroom with a stall, a line of knobs for towels, a large message board, and a separate telephone room reflect his penchant for efficiency and hope for community.

Mick takes me to see the one available cabin, explaining that "the cabins have electricity but no water. You need an invitation to go to someone's cabin. Everyone's independent, but sometimes we eat together during the mandatory weekly house meeting. It's all wood heat and you have to chop your own; the stack is by the front door. The phone is shared. Notes for each other can be left on the message board. Kind of like having eight roommates, but spread out across the property. Well, here it is."

We reach the cabin and I have stopped listening. A demure window-speckled wooden house snuggled under the arm of an oak seems to wink at me. I stop walking, stunned with love. Mick, maybe sensing this, ambles off. "Take your time. I'll be in the kitchen." Bright, clean, cozy. A skylight over the bed loft. This ten-square-foot cabin in its copse of oaks bewitches. I sit alone there awhile, watching leaf shadows animate the wood floor, absorbing the silence into every pore. Beginning, already, a romance with this place.

We stay until noon and meet the rest of the community. I shake hands on it and we ramble back to Eugene, mellow with fresh air, scenery, and success, where I treat an appeased Gabriel and Sophia to lunch at a chic café.

Migration

Like a bird honing in on its nesting ground, I searched
the map for someplace far enough from traffic, but close
to Eugene where I still had friends and work. Why did I
feel compelled to live out of town? There's mystery in my
motives, but a couple are discernible to me: recovery and
renewal.

Recovery: My college sweetheart (now a jet-set archi-
tect) and I (a hopeless bumpkin) have finally decided to
give up our dream of a life together. I couldn't live in
motels and Alex, raised in Hong Kong, developed his
image of rural living from the film *Deliverance*. The country mouse and the city mouse. We still like each other
better than most anyone else on the planet, which is
why it took us over ten years of a mostly long-distance
relationship to face the obvious, and why I am now walking around like a ghost that misplaced the body it was
borrowing.

And this is my attempt to wriggle freer from a consumer lifestyle that I know to be harmful to nature and
to our nature. Granted I was raised Catholic and prone
to guilt, but what are we here for, if not to live ethically
and to care for what we love? It has become harder for

11

me to blithely get in a car and spew toxins into one of the elements that make life possible. To mindlessly turn on the house lights and imagine salmon stopped by hydroelectric dams, lost in the confusion of waters behind them—or to throw away plastic packaging and recall rainforests ravaged for petroleum and of the unique cultures destroyed in the bulldozers' wake. Not that I feel this way often, but often enough to give pause.

Renewal: I am beckoned to renew a childhood connection with soil and sky that once offered me solace. I grew up outside Chicago, in an area with large yards and immense trees, where the variable Midwest weather enamored me. As a quiet middle child in a boisterous family, I'd climb our willow tree with a book and a blanket for whole afternoons, squeezed and wrung out by big-shouldered clouds. Or spend afternoons lying on the grass in our yard beneath a willow tree, watching clouds pass overhead and filling with bliss and a gratitude for this "secret," this capacity to let nature open me to joy.

It was a kind of soul sustenance for which I've felt hungry for the past few years. So I've come to Siesta Lane to at least put my ear to the ground and listen for what is wanted by and from me, and to begin to respond.

Willow

From the first time
one beckoned I was taken
by the tree people
into their fold.

Scraped, dizzy, bruised
but given over
up and up
I knew my real home
was here.

Where wind curled branches
birds built forts and no one
on Earth could find me.

Gramma held
her veined hands beneath
a floured apron called
and called to me.

Vagrant girl pressed against bark
caressed by yellow fingers
catkins in hair
heart of sky.

Entrance

The nine of us have arrived separately to these eight acres on Siesta Lane in Rainbow Valley and squirreled our extra belongings away into the barn. We each have a rustic cabin and share the main house. Mick is a botanist who thrives outside; Sara a grad student with few possessions; Raul and Luke printmakers who like the low rent; Jack and Rita social workers here for the sweet setting; and Paul and his daughter Natalie are happy for a country place to grow food. Our secondary reasons are more subtle: artistic inspiration, intimacy with nature, a more basic lifestyle, human community. Living here will be an exercise in simplicity. The largest cabin, or rather, round wooden yurt, is sixteen feet in diameter and it's a long way to the storage barn, especially in Oregon's rain. We will learn to get by on less, and to both curse and welcome the sparseness. The asceticism will leave us open to watch for grouse and grosbeak, I hope, and as the human world shrinks, the natural world will pick up its bags and move to the front of the line.

Sara, in shorts and a sweatband on her way to ultimate Frisbee practice and whom I had at first sight recognized as friend potential, comes out to help me move my red satin armchair up the hill. "Wish I could help more, but I'm already late." Her cropped golden hair swishes her shoulders as she speeds off. This flusters me and I bump the chair against the doorframe fitting it in. I had been romanticizing the folks here as enlightened souls, sitting around all day meditating on the dandelions. So maybe this place isn't a magic charm after all and everyone here is even busier because they have to drive twenty minutes to their other lives? I will need to be diligent about idleness, I think, and sit down hard on my worn armchair as it releases a wheezy sigh.

It has occurred to me that I rush through the days with my arms full of projects—usually on my way to pick up another. Carrying my weight, proving my worth. I have been out in the world doing my life. But not being it. This is what happens when you go to the woods to observe. Time stops long enough for thought and feeling to catch up. A friend did the same without going into the woods. He stayed home for three days and did nothing. Just sat. In the dark. Alone. Just to see what would happen, what it would be like.

Such a call to introspection sinks me like an anchor into a deep lake that is at once seductive and terrifying. Feelings catch up to me when I slow down—my pact with mortality for one, but also the deep riches of solitude. I become humbled and grounded, the way a pond clears when the water settles. I crave this clarity and depth, to be shaken and then stilled. But first, shaken.

A lonely howl echoes through inner chambers and for a few minutes I consider bagging it. I haven't signed anything yet and Quin could find someone else. Part of me knows that it would be easy to go back to my busy familiar life where I am marginally successful and known to others.

But not to myself.

I will stay.

And then will my time here be a ripening, or a kind of dying? Will I, like Rip Van Winkle, fall into a stupor, or will I deepen my current from gathering in, listening and watching, watching.

I get up for another load as the sun ignites the oak leaves against my cabin window and realize there is no lock on my door.

Community

When a cabin opens up at Siesta Lane, the owner finds a new tenant with the current tenants' approval. Over the past two years, Quin had assembled our particular collection of eccentrics with me as the lucky last taker. "Just try to get people that are halfway normal," Mick, one of the early renters, had told him. And Quin had. But fortunately, only halfway.

In the coming weeks, we'll learn that Paul needs to sing loudly when he cooks; Mick's plant specimens are growing on the window ledge; Raul, a Shoshone Indian, wants to keep the deer brains in the freezer, which he uses to cure his drum leather.

When you live with eight other people miles from town you learn to accept your differences. The surrounding countryside has a way of putting people in relief against it. You recognize yourselves as the same species. And stick together. Although mostly independent, sharing the main house makes us interact daily. We have to divvy up chores and utility costs. We all go into "the world" to work and play, but most nights folks choose to hang out at Siesta Lane. Sometimes we cook together or have an event. One night a month we have a gathering that we call Art's Night Out. We make cookies and present pieces from our art or work. Somehow through cracks in the walls of our separate egos, we come to know something of each other.

Jack and Rita are newly married and covetous of their privacy, going to all lengths to be alone in the main house at mealtime. Because of our erratic schedules, this causes them no small stress. Unpredictably between 5:30 a.m. and midnight, Siestans will pop into the main house for a snack or cup of tea while Jack and Rita are trying to sneak in a romantic meal together.

We suspect Jack of trying to live up to Garrison Keillor's description of a Minnesotan. "Has he said anything today?" "Nope." "Well, write it down if he does." The only thing we really know about Jack is his extreme patience.

We know this because his new bride, Rita, talented and restless, keeps uprooting their lives. She wants to be a nature educator so they move to Boston to go to the best school. Before finishing, she decides she really wants to be a Waldorf school teacher so they move to Eugene, but she wants to be out of town so they move in with us but since then she has decided that the best place to learn her craft would be England, so they're going there next year.

Jack, who works as a contractor, finds work wherever they go. He puts a hand on Rita's knee, eats pancakes, and reads the paper.

Mick is a stalk of wheat. Moppy blond hair blowing around his blade of a frame. He's quiet, responsible as a patrol guard, and has the distinction in our household of only taking up one quarter of one shelf in one of the four refrigerators. We think Mick exists on peanut butter and jelly and expect that he's secretly involved in a study for the FDA on just what that diet does to a person.

Amy Minato

If we need to know what any plant is, we ask Mick. One night he gives us lucid descriptions and titles of the parts of grass, which I promptly forget. We gather around the table, heads bent as in worship, sporting headlamps and hand lenses, peering down at bits of scattered native grass. The task is to decide if the ligules at the base of the leaf sheath are hairy or membranous (if there at all) and whether the appendage-like auricles below them (if there at all) are short and stubby or resembled claws. Of course, this is after we determine whether the seedhead is a spike or a panicle.

Sara, at twenty-four, is the baby and the angel of the group: golden-haired, blue-eyed, with skin like a peach. She likes to bake and knit, but also plays soccer like a Viking and gives lectures to graduate anthropology classes at the university. We'll be sitting around shelling beans from the garden and Sara will timidly startle us with some revelation from her past, which we had imagined culminated in the high school prom. "That happened when I was studying Barucan weaving in a remote Indian village in Costa Rica." Or, when asked if she'd read Meridel Le Seuer, "Oh, yes, my mother is a good friend of hers and is writing her biography."

We trip over ourselves to get Sara to smile (which isn't hard) because she has such nice teeth. We perform embarrassing antics just to hear her bright laugh, quick as a splatter of rain.

Raul, who is maybe five feet, five inches tall, has extraordinary long black hair that he keeps impeccably clean and tightly braided. He has the smooth brown skin of his heritage and a temper even as an ironing board. The mysterious paintings he makes of spirit plants, life under the soil, and eagle people stun and silence us. We believe we are in the presence of a great artist and sage. Only when he tells us that his wife put all his belongings outside his house (which is the Shoshone style of divorce); when his parents call asking for "Ralph," which is his given name; or when he mentions his son in a sad voice do we concede that Raul, though immensely talented, is human, searching like all of us for some reconciliation within himself on this particular quilt patch of land we call Siesta Lane.

And there's Luke, whose last name means "eagle" in Spanish, coming to us from the pizzazz of Los Angeles. Luke makes elaborate colorful prints and thick espresso coffee, and wears silk shirts. His girlfriend, Yoko, is from Japan and comes to about Luke's navel. She, too, is an artist. "Excuse me, Amy. May I photograph your boots?" I realize she is referring to the boots that have been left shamefully outside my door through monsoon weather and now have a spider web in them. I guess she's never seen boots like those in Japan. "Okay," I say sheepishly, "only please don't put my name in the title."

Luke, who has curly black hair and is at least eleven feet tall, has the softest voice I have ever heard. He could get a job lulling children to sleep. He stands out in our household for knowing exactly how to cook a perfectly fried egg, which he has for breakfast every morning. He is the one we turn to when we need something written in exquisite, artsy handwriting.

Although his daughter goes to school in Eugene, Paul, who hails from farm stock, requires country living and has been at Siesta Lane the longest. Paul has a deep voice and a leathered face but an elfin grin. He knows how to fix anything in the house and how to farm a parking lot. The sprouts he keeps on the windowsill never rot or dry out. His daughter shines with his good care. Pre-teen Natalie has the lithe, sturdy quality of a foal, and walnut eyes that pulse with delight and ferocity. She knits, grows beans, builds fires. Paul works and works, worries about rent, and ages fast. They are an inseparable enigma, a cameo of pioneer survival that makes the rest of us feel as if we are playing at life.

And that leaves me. A single woman with ninety-eight part-time jobs trying to figure out how she connects to the greater forces of the universe and to the tiny earwigs that hide in the curled-up seed heads of Queen Anne's lace.

Incubation

Nearly hidden in the thick foliage, a long grey nest of duff and dead grass hangs like an old sock in a clump of hawthorn shrubs along Siesta Lane. Good trick the bushtit has, hiding her home amid thorns. Determined to make a study of the nature here, I wait near it all morning with my notepad and sketchbook, watching for signs of life.

Male bushtits, small acrobatic birds with long tails and feisty attitudes, will build several decoy nests near their occupied one to attract mates and to deter predators. I'm patient, hoping this one is the real thing.

One summer I shared a house with an environmental artist who designed "dwellings" from found material on the land. Our backyard was decorated with shelters that ranged from miniature to tent-sized. A circle of stones marked a meditation spot, branches made a teepee for afternoon naps, and grapevines grew around ropes creating what Carolyn referred to as a "menstrual hut" where we were to rest and reflect during our periods. Our neighbor, watching Carolyn gather dead grass for mats, remarked in a quiet voice, "Maybe she needs a mate." But I interpret Carolyn's art as a gesture toward making a home in the world. For Carolyn, a dwelling is a green place that smells like dirt; for me, a small cabin on an oak-studded hill; for the bushtit it's a tenuous tube on a thin branch.

Suddenly the sock begins to quake and a flurry of tiny brown birds flies out from the nearly invisible hole in the front. Nerves tingling, I note their presence as a good omen on this current spiritual quest—my new attempt at creating a home on the land, hoping one day I too will release such exuberance.

July Moss soft on the oak limbs that dance outside the window, ripe plum on the sill. To sit with myself, a pen, and time is a triangle of apprehension. I let the feeling be. I keep the pen in hand. The oaks darken. We begin.

Freaks

I explain to my conservative mother that I am living in a "community," not a "commune."

She is familiar with Moonies, orange-clad Rajneeshies from central Oregon, and Patty Hearst, and is not easily mollified.

"Honey, it's not one of those cults is it?"

"No, Mom. Everyone's really nice."

"They're always nice in the beginning. That's how they suck you in."

"Really, Mom. We hardly even see each other."

"Just don't let them brainwash you."

It gives folks pause when they learn that none of our cabins have bathrooms or kitchens or water at all. Walking a hundred yards outside in total darkness to the bathroom strikes them as barbaric.

Siesta Lane

When my brother visits he's initially shocked by our occasional use of outdoor plumbing, but ends up appreciating the excuse to get outside and see the stars.

We play it up when he comes, knowing he's skeptical, and maybe even enlisted as a spy by my family to see if I am living in some drugged-out hippie commune or New Age cult center.

So in place of our usual ragtag assortment of phone messages and chore lists on the chalkboard I write the following and leave it up for him to find. For many years, he will not be convinced I was kidding.

Daily Schedule

7:00 a.m.	Sun chant/ritual cleansing with Guru
7:30 a.m.	MWF Mantra TTH Colonics
8:30 a.m.	Macrobiotic breakfast
9:00 to 5:00 p.m.	Work for capitalist pig-dogs
6:00 p.m.	Vegan dinner
7:30 p.m.	Sunset drumming circle
8:30 p.m.	Tarot reading (optional)
9:30 p.m.	Aromatherapy massage
11:00 p.m.	Group howl

Everyone: Amy's brother will be here this weekend. Please wear clothes.

Writing

In my loft on Siesta Lane I lie on my back in a cliché of laziness, watching a dense cloud through the skylight drift steadily west over oak and red pine, sponging up the sunset. My cloud chart, with pictures and interpretations of various cirrus, stratus, and cumulus clouds, tells me this means rain soon. I am relieved there is some practical purpose for my vigil.

What is lost when I deny myself cloud-gazing? When my life trots with assignments, my mind hones to production and profit and forgets about rosehips blushing above their green collars, the creak of passing squirrels, hawk feathers disintegrating into mud. What if I can learn to turn off the monkey mind and tune this swagger of bones to insect frenzy and cloud swirl and the secret stretch of xylem and phloem?

What I hope to excavate from here, on my knoll in Rainbow Valley, cradled in wood heat, is what in my nature has been atrophied by modern culture, what qualities weakened, and to exercise those forgotten sinews. To stumble toward a health of the soul, and tell about it.

I go out to play with Ceres, an old, limping, half-blind, mottled grey, white, and black Australian Shepherd who belongs to the owner and gets to live on the property. One of Ceres's ears flops to the side, as if she is listening to sounds outside of herself with one ear, inside with the other.

She sees me search the brush, knows I am looking for a stick. Her bump of a tail wags furiously, her ears peak, her whole body quivers and snakes. I find a thick, sturdy stick, and whizz it across the grass. Ceres is off before it leaves my hand, bounds through fescue and poison oak, finds the stick, and waddles back.

Pretty soon the stick is a crumble of wood chips in her jaw. Only then does she remember me. I throw a second stick, this one part of a gnarly oak branch. Ceres returns with it, closer this time, a yard away. Hesitates. What is that look she gives me? Wariness? Apology? She chomps the stick, not seeing me anymore. This one will take longer to destroy, and it's clear the game is over. I trundle back to my writing, to my own twisted stick, which I gnaw and gnaw.

The Garden

The geeks and mystics are feuding over the garden. Those with a practical bent are into maximum yields of veggies, using rows and measurements to achieve them. Those with an inclination toward spiritualism want to plant flowers and medicinal herbs in circles by the light of the moon.

I am sitting on the fence, good middle child that I am, trying to mediate: "What about curved rows? Maybe the new moon is this weekend. Should we split the garden?"

That's what we do. One half becomes a spiral of sunflowers and rosemary, the other a veggie factory, a conveyor belt of beans, corn, potatoes. Together they nourish body and soul. Our yin-yang garden.

Impulsive people are, it seems, more likely to be outcasts—shadows lurking around the edges of society— crazy artists. Western culture values, *requires* order, efficiency, organization. But it is nonlinear folk who keep the spirit alive, honor mystery and the irrefutable bent of everything in the universe toward chaos.

It's a daily struggle for artists to survive in the modern world. We forget stuff, change plans, get places late, fall in love. We are called flakey, spacey, irrational, and unreliable.

The differences among Siestans surface every once in awhile—with cooking, fire-building, and cleaning styles. Some work from the heart and others from the head, with the more spontaneous among us grating on the organizers. "Can't you leave the cleaning supplies in the order you found them?"

"I like variety."

"The schedule says you're supposed to take out the garbage on Fridays, not when you feel like it."

"Chill out, it's not full yet."

Clearly our community needs this diversity—those who follow directions, and others who improvise. If equally valued, these approaches complement. If disrespected, they polarize. I'm beginning to wonder if these styles represent profound attitudes toward the earth. Do we work in unpredictable harmony with it? Or introduce our own structures and timelines? Is it possible to do a little of both, like a musical score that deviates from its baseline rhythm, a blueprint fleshed out with surprising details, or a story, receptive to its audience, that changes in the telling?

In the meantime, steadied and enlivened by the forces of both sun and moon and surrounded by agents of each, we tug and give and surrender to the influence of both order and magic on the fabric we make of our lives.

Gratitude

Yoko's at it again. Hoisting a camera the size of a mail-box in delicate hands, squatting beside pea blossoms, hovering at the utensil drawer, angling the camera along rain-streaked windows. Her face these days is a big lens surrounded by hair. You might be furrowing your brow over the newspaper and hear a click. Immortalized.

At first it's disconcerting; then we join in. Scavenging for plant pots in the barn, Mick and I notice Yoko with camera poised. We huddle behind her and peer down onto the shadow-striated floor. "That *is* kind of cool," Mick admits. We keep looking at whip marks of light on dusty planks long after Yoko has moved on to dented buckets, wondering that we never noticed them before.

The plethora of "goods" in our society can sometimes dull me to nuances of sky, hues in an apple, a friend's handwriting. For presents these days, I give handmade gifts, services, or experiences—a knit hat, a massage, a nature walk.

At an eighth birthday party I watch my friend Gloria, a political refugee from Guatemala, wince as her daughter

Maya tears the wrapping off gift after gift, discarding each as soon as it is open.

"When I was a girl," Gloria says softly, "I would carefully unwrap my presents so not to tear the beautiful paper. I would keep it in a drawer to look at. How happy it made me!

"My daughter is never that happy." Gloria flutters her hands like small ducks rising off a pond and resettling. "She is never satisfied! She always wants something newer, better, something she saw on TV. . . .

"Yes, we are safer here, and healthier even, but we did lose something coming to this country. We lost contentment.

"Here, we are guaranteed the 'pursuit' of happiness. And you know what? I am tired of chasing it."

August   Beetle climbing the phone cord, waving long antennae in that comic bug look. Those faces that never change expression, cartoon eyes, and bodies too big for their legs. Then the frenetic buzz of a fly against the windows, zippering the glass to get out. Nothing in her genetic memory has prepared her for her inability to go there, to the place she sees before her, that branch, that bark.

Mortality

Yellowjackets crazed with appetite gorge on Raul's half-eaten chicken leg, meticulously chaw off pieces bigger than themselves with their front mandibles, hover, and stagger off.

Maybe we fear insects because there are so *many* of them, because they will outlast us, because we can hardly see them, because we blithely do the most dastardly things to them, and because they are finally, in this feast of life and death, what eats us. A box elder bug struts across the picnic table like a bored buyer at a furniture sale.

Specimens

We study
the geometric messages
encoded on the back
of the box elder bug.

Hieroglyphics
carried on a carapace
black as a flake
of obsidian.

Like a magician
or a buddha, it holds
its big toenail of a body up
on a trivet of legs
thin as lashes.

And in absolute silence
on the maple leaf
one antenna
strokes the air, conducting
a private symphony or tuning
into our vibrations
listening back.

Insects here become a different presence—especially the hypnotic buzz of crickets. They sing by rubbing wings together, which is a bit like singing and clapping in the same motion, and listen to the vibrations through tympanums on their legs. Their chant reflects a collective soul, I believe, different from the individual chortles of birds, as if they are in a trance state ready to elevate into a higher incarnation.

Still, we are experiencing an outbreak of yellow jackets and I have yet to make peace with them. I've grown accustomed to poison oak, red leaves flagging the grass, how it clusters in constellations across my skin, stinging me to life—ferocious, unrelenting. Cause of the most common, and perhaps most ancient, allergy on earth, a quarter ounce of its urushiol oil is enough to give everyone on the planet a rash. Poison oak says, "See me if you want. My hands are red. Go on, ignore me. I'll talk to you later." I'm learning to accept the seeming inanity of its outbursts, as bail for my wanton wanderings. Those small itches and rashes needle me awake, remind me that I have skin, that I live in flesh. Anyway, it can't chase me.

But the yellowjackets! Maybe it's the sound they make, disturbing me in my meditations. They say, "Here, here, now here, turn, swat, worry, mind, be wary." Late summer is yellowjacket time. The hotter and drier, the meaner they get. They're dying and they're pissed. Yellowjackets crawl over my food, into my mug of tea. They are invasive, pervasive, swirling in a frenzy of appetite, especially aggressive if they smell meat. It's weird to think of insects as carnivorous, but these are, and if you kill one, it gives off an alarm scent that harkens the swarm.

All reason and Buddhist leanings aside, I roll up a newspaper. Aim and swat, aim and swat, aim and swat, circling, stalking, wondering at my own ferocity, the terrible match we make.

Carpenter ants inhabit Sara's cabin, which Quin decides is no big deal. "Live and let live," he says. "The carpenter ants have been gnawing on that cabin off and on for years. I've seen a lot of carpenter ants, and I've never seen a structure fall down because of them."

Sara's cabin is lovely and she doesn't want to move. Her room hosts weavings and photos, quilts and candles, bright clothes and books, and windows to the surrounding woods.

Amy Minato

Yet the ants chomp her roof and peek through the cracks, wings glistening in the moonlight, antennae curved and waving. On a bad night, she fears they will drop into her eyes and hair. She dreams of bones and dirt. In the brood chamber of her ceiling, ant larvae fatten in silken pupal cases, fed on grubs brought by infertile female workers. While she sleeps, as many as two thousand worker ants cooperatively chew up beams over her head and spit sawdust into their inscribed galleries. Sometimes, she can't sleep. When Sara knocks on the ceiling, the ants reply with a rustling sound like cellophane crinkling.

I observe a roadkill raccoon along the road to Siesta Lane. One day it is suddenly there, whole and seemingly unharmed. In the next few weeks I follow its slow decay. A skeleton emerges from the grizzled fur, blood congeals, grubs set up house. At first I walk on the other side of the road, looking off into the trees or anywhere. But at last I kneel beside it, honoring its life, curious about its anatomy, sorry for its death.

This raccoon may have been the one that I saw "washing its hands" at a nearby pond. Later I learned that it was actually feeling its prey under the murky water, to sense what it was about to eat. Water increases the sensitivity of their paws. Paws that can lift lids, unscrew hinges, remove container seals. Now they were curled up and stiff, still eerily human.

On this homestead we have but a few illusions to string like tinsel onto the ramshackle scene. We can caulk the cracks and dress in silk, but the carpenter ants gnaw all night in the beams and yellowjackets pinch us awake. At least we are relieved of the effort to *pretend* that our skin is not sloughing off, that insects are not feasting on us and our houses, and that decay will not eventually chaw into pulp this home, our bodies, these fictions.

Circulation

I don't remember owning anything as a child, or thinking that was a possibility. Sharing was right up there with eating and breathing. My parents had a modest income, seven kids, and huge school payments. A Catholic education, in their opinion, came before stuff. And Catholic training reinforced the idea that people are connected, that what happens to a prisoner across the world is our business. Later, I expanded the web to include nature. Eroded soil anywhere, I concluded, concerns me.

We shared rooms, baths, beds, and our parents' laps. There were no names written on things. Games were kept in a closet, books on the shelf. All toys and clothes were passed along in one great communal river. In some ways we were one being, like those quaking aspen tree stands with identical DNA that share the same roots, considered to be one of the largest organisms in the world. To this day my mother can't keep our names straight.

Although I moaned sometimes about wearing hats and coats that looked like Halloween costumes, and drinking powdered skim milk, I survived to tell about it. We were rarely sick and never cold or hungry. And for our fashion-deficient household, school uniforms were a godsend. The fifteen-year span between eldest and youngest wasn't enough for things to come back in style. We finished everything on our plates and leftovers were efficiently wrapped in reusable containers.

Last year, when our parents finally ripped out the thirty-year-old shag carpet, it was an archaeology dig of our youth—crayons, marbles, buttons, rings. Vacuums, it seemed, hadn't penetrated the depth of this neon green lawn on which we'd crawled, played cards, wrestled, and shed myriad skin cells. In every holiday photo, this acrylic foundation had underscored the pyramid of enlarging children in new combinations of the same old clothes. And though we'd groaned for years about how ugly it was, it had acquired what the Japanese call "the patina of human use," and in a way we hated to see it go.

Not owning anything usurped power struggles over "stuff." "Mine" may as well have been a foreign word in my family. Our identities weren't wrapped up in what we had but who we were, what we did. Like cedar waxwings that line up along a mountain ash branch and carefully pass berries down the row, we took care of something for the next youngest, because it had been taken care of for us.

As nothing was discarded, there was a sense of abundance, and anyway, things gained significance as hand-me-downs. I remember being on tiptoe, eyes wide with worship, my tilted chin barely resting on the ping-pong table on which blue fabric was laid to cut, watching my older sister sew a dress for a dance. Later when I wore it, although too short and too loose in the chest, it was tangible proof that I had arrived somewhere. And passing it on meant I was still moving.

The times when I've felt more possessive and acquisitive have coincided with my periods of insecurity and struggle. We need, once in awhile, to cling to objects, infuse them with meaning and let them reflect us. But such a focus can become a quicksand and a stagnation. When life changes or natural disaster clears our lives, it is freeing to emerge, even unwillingly, from the clutch of possessions. To be more motion than material.

I've come to feel this way about the earth as well. To celebrate the fact that each atom in my body exploded from a star, that some molecule in every gulp of water has circulated through a dinosaur, watered a palm tree, evaporated off Sappho. It seems clear and even freeing that I am to borrow only what I need and caretake the rest for future people, bromes, and arthropods.

Altar

I wander the earthways near my cabin and instinctively gather mementos to borrow: a heart-shaped stone, red fungal lips on a mossy twig, a swath of fox fur. I place these in my pockets and empty them onto a low wooden table at home beside shells and cones, seeds and dried flowers. People bring the outside in, it seems, to keep connected. Plants. Shells. Rocks. Flowers. Natural items decorate our homes, and windows provide living pictures. Many of us have not grown fully accustomed to the sterility of life indoors.

Such collections predate science, and were crucial to our survival. Early humans must have gathered plants and experimented with their use for food and medicine, finding that chewing on willow bark helped with pain, juniper berries staved off colds, rose hips cured scurvy. Although curious about practical applications,

Siesta Lane

I'm prompted more by poetic possibilities, spiritual nourishment, and a naturalist's curiosity.

Throughout the week I pause over my collection, appreciating the new arrangements of color and texture. Small tokens feed me intimately, in a way landscape can't. As if nature is too much, sometimes, to be in. Here I absorb a bit at a time, giving clear attention to a flicker feather or wasp gall, without distraction from a passing hawk.

Some nights I turn off the lights, sit on the floor, and meditate. Candle glow sweeps fiddleheads and dragonfly casings. In the night's trough of silence, my "found" tarot suggests nuances of the human experience: transience, hope, vitality, death. Breath slows. Spirit fills.

In a past era I might have been killed for this activity. Women and nature, especially, add up to a powerful combination. Female knowledge of and connection to the earth have threatened male authority for centuries. Maybe men were jealous and frightened of women's ability to pop human beings right out of our bodies. Whatever the seeming threat, during the European witch hunts of the sixteenth and seventeenth centuries, thousands of women were burned at the stake for less. For providing an herbal remedy to a laboring woman, maybe (who was expected to suffer in payment for Eve's sin), or for daring to walk alone at night. Burned. At the stake.

After a few days I return most items to the land for the inhabitants to use in their fashion, or for the next inquisitive visitor. Some treasures I hold onto for a day, a week, years—their whispered tales still spinning. A spider casing stays with me. I look and look at this delicate shell, wondering how the spider pulled her thin legs out of that translucent skin. Spiders have no internal skeleton, so they have to molt in order to grow. During molting, until their new skin hardens, they can't move their soft bodies and are extremely vulnerable to predators. Why does this intrigue me?

Day after day I resist returning the casing to the land. The light prisms off it, the lack of spider inside giving it a spectral integrity. "Look," it says to me, "I was once useful but knew when to slough off. Commend me." I do. And notice the itch of tired personas urging me to move on. It's unsettling to let go, to not become fixated with familiarity. A practice for the larger yieldings of change and loss.

Through the simple ritual of my revolving altar I learn to trust life's tilting kaleidoscope, the heart's dynamic journey.

Sustenance

To survive I work several jobs. As a published poet I teach writing in public schools for two-week stints as a visiting artist. Mostly I help edit *Skipping Stones*, a nonprofit journal by kids for kids, focusing on cultures and nature, which I co-founded with a man from India. Kids from around the world create it, really. We just arrange the words and pictures. It started as a gesture toward world peace through communication and an appreciation of diversity. We wanted to empower children, offer them hope and a sense of justice, and publish their inimitable writing. This, like most of my life work, though greatly rewarding, involves tedious hours and offers abysmal pay.

Sometimes I answer phones in the office (i.e., garage) of a friend who hopes to fool clients into thinking he's big time. Having a secretary, he says, shows you've made it.

I substitute at a bakery—a place where the bakers sing while they knead bread together at a long wooden table. We listen to music, smell yeast, taste bread, feel dough, and watch hands. Sometime I speculate that living in town, I could work there every day. But it's not what I want right now.

Amy Minato

I get by, mostly, on these sporadic funds, praying that my teeth and body hold out. None of us have much money so we live cheaply. Firewood and utility bills split nine ways are meager. Shared and garden food goes a long way.

Luke, Raul, Sara, and Rita go to grad school; Paul and Jack work construction; Mick's the only professional, but serving The Nature Conservancy comes close to a labor of love. Basically, we've achieved economic parity. No coveting or condescension in this neighborhood. Besides, there's no room in our cabins for riches, and stuff rots in the barn.

But are we the little pigs building our houses of straw? The crickets who fiddle away all summer while the mice store food? College friends write me about their savings plans and retirement funds. I tell them how red the fox's fur looks in the afternoon light. They think that I will end up a bag lady unless they can fix me up with a rich guy. Local ones try.

I go on one date with a law student and feel my soul slump under the chair. He orders steak and martinis and talks about estate law. I tell him how close I came to entering law school—back in one of my more practical phases—passing tests and getting accepted and almost signing up. I almost signed on but then it occurred to me that I hated arguing, and was so easily persuaded to empathize with an opposing view. I pictured myself in the courtroom, walking over to the other lawyers, acknowledging their points, and withdrawing from my client's case. Law school, as would this law student, receded from my future like day before the chase of night.

To ever have a family I'll need to buck up. Not everyone has this luxury of a hiatus in midlife. I'm told to "figure out what I want to be when I grow up," but puzzle over how to gracefully fit my new person into the same old world.

Making a Living

One morning I wake to the clatter of a squirrel acrobating above my head. The five toes on a sqiurrel's back feet and unique ability to swivel its ankles make it a star circus act. This furry sentinel stops to peruse me through the skylight while oak branches knock on my roof. My bed is covered in papers that I fell asleep reading for *Skipping Stones*, and there's no way I can make the board meeting that starts in town in five minutes. The squirrel seems to realize my dilemma, and is waiting to see what I will do, thinking, no doubt, that I'm about to put on a good show. A weeping or cussing fit maybe. A comic attempt to dress in two seconds. Instead I stare back into his black eye and hold a silent conference between my yin and yang.

Juggling several jobs and projects is not conducive to a contemplative life. But I like everything I do. The magazine pays nothing but the kids' writing jazzes me, and it's my offspring. Answering phones for a computer security business is inane and meaningless. But it's easy and pays shockingly well. The bakery hours are too darn early. But the smell of baking bread, the smooth, cool feel of dough ...

Amy Minato

I realize that at this point I should call my rational friends, my parents or those college peers with the security funds, for some common-sense advice. I should flip through investment magazines or go to law school. But what do I do? O child lacking reason? I look outside at trees.

The banter goes on in my head until the call of the oaks wins out. If I never have time to be with them, to be embraced by their branches, lost in reverie of flickering light and leaf, what's my life worth? I've missed the meeting, but I'm here.

Knowing that I am crazy and doomed but saved somehow, I take a leave of absence from the magazine, quit the bakery and fake secretary job, and decide to live off freelance writing and my occasional artist stints in the schools. I resolve to apply for fellowships and plant a lot of vegetables.

Immediately, I accrue unexpected payments. My car breaks down and my cat, Quixote, gets worms. The magazine is short on funds and can't pay me for my final month of work. My dreams of simplifying turn to nightmares, literally, of indigence and shame—running down the street naked. Breathing deep and maintaining my resolve, I ask Sara if we can car-share for awhile, cut out all extra expenses, and sell an antique chair that I find at a yard sale to get by. I can't do this forever, and if I had children I couldn't do this at all. But for now I don't know how not to.

Devotions

My daily walk often leads to a small, forested park on a hill, a green hump above gold fields, a place I've come to think of as my temple. Unfortunately a neighbor has taken to putting up signs that say, "Keep out! Private

property." I'm miffed and hurt and ask Mick about it later.

"They just think it's theirs 'cuz they've been using it for so long. It's not. It's a public park that just doesn't have a sign on it. You can go there whenever you want."

So I do. And, of course, one day a fellow in a very large pickup happens by and stops. "Private property, lady," he says.

"Well, actually," I clear my constricting throat, "it's not. It belongs to the county."

"You tryin' to tell me," his ice blue eyes check out my boots and backpack, "where my property is?"

I fumble into my bag and pull out a map. "Here, see." I point to a green triangle on the otherwise white grid. "It's public."

Without looking at the map, he mutters and drives away. A large dog beside him growls and bares yellow teeth as they pass.

I breathe deep and keep walking. Dew-speckled trillium, ginger, and bleeding hearts twinkle along the path. Mick has asked that we look out for unknown plants here. He thinks there may be rare ones in this undisturbed place.

Inside the butterscotch cup of a salsify, a crab spider shingles itself to a petal, waiting for prey. Crab spiders, flat with crablike legs, can move sideways or backwards and change color to match each host flower. An insect cruising into a calyx for some nectar is in for a nasty surprise.

A circular maidenhair fern crowns a fawn lily beside a patch of fringe cup next to a reddish, triangular-leaved plant that I haven't seen before. As I bend closer, a twig snaps near me. A deer, maybe! I look up from fingering the coiled fiddleheads of ferns and lock eyes with ... a cow, munching determinedly on the tender young plants of the forest floor. Dismayed, I shoo it downhill to the denuded pasture where it belongs.

That's why the pickup guy wanted me gone, I realize—so I wouldn't discover that his cow is sullying the creek with waste and crushing fragile native plants with its hooves on this rare patch of public land.

I know that I will have to call attention to this, somehow, to alert authorities, to be the "meanie." Because when we side with ecological health over someone's personal interest, we are seen as inhumane—as if all of our children's future is not dependent on the state of the earth they inherit. As one crusty activist put it, "Environmentalists may not make easy neighbors, but we make excellent ancestors."

At the top of the hill there's a clearing where you can see most of Rainbow Valley, a rolling patchwork of trees, scalloped sky, and horse farms. I bow to each direction, thankful for health of mind, body, spirit, emotion. Afterward chickadees flit behind the tinsel of lichen on the oaks as I sit for a long time, listening.

At dusk I wander down, feeling raw and protective. Because I know these trees and their squirrels, these ferns and wildflowers, I care about them. The burden of responsibility balances the pleasures of love.

And what in nature is not threatened today? It's crazy to become fond of a natural world gasping from our assaults. How much easier to watch it on film and in museums, or to casually visit wild places leaving no heartstrings attached. Then when you hear about their destruction by pollution or resource extraction it causes just a small sadness, instead of fierce loss.

So it's understandable that many of us close the shades and turn on the TV, sheltering ourselves from the devastation of this planet and our complicity in it. But how does it affect us, especially kids, to stay inside? A parade

of recent books bemoans
the condition of our techno-
savvy, nature-impoverished
children, correlating the
rise of ADD, obesity, and
childhood depression with
today's indoor culture.

The more detached we
are, I believe, the more mis-
erable we become. A terrible loneliness grows in us,
a loneliness for the earth.
No consumer goods or even
friendship can assuage this
grief. There is no substitute
for direct experience, for
sun on our face and soil in
our boots. We need to risk a genuine, daily relationship
with place. The only remedy is to take that precarious
step outside.

September

Wicker chair with a broken straw seat, a lantern with smudged hand marks, small cracked table covered with low shells and stones and feathers, a candle burned down low. A tiger cat curls graciously around the lantern, batting at moths, volleying them between its soft, lethal paws. I'm amazed at how quiet and sure a cat walks without looking, tripping, or knocking anything over. Oh, to be so careful in my own ways.

Accessibility

The cabins don't have hook-ups, so the nine of us share
one phone, which is always in use. We line up patiently
outside the narrow, closed door, trying not to overhear
someone else's conversation. The phone room is not
unlike a confessional, cramped and dark, where you talk
to someone you can't see. With eight adults here, when
the phone rings, the odds are that the call you answer
won't be for you, and whomever the call is for will be
in their cabin. So to appease the caller and because
it's a link to the world, you end up in conversation. I
catch myself telling near strangers the menial sins of
the week.

The phone is our main link to the outside world,
which might as well be another planet. That and the
mail, which arrives to 28577 Siesta Lane as if there were
twenty thousand instead of six houses on our road.

Most of the time when someone calls me the line is
busy, no one's around, or I'm in my cabin or otherwise
hard to find. It's rare to actually receive a phone call.
Mostly I get hard-to-decipher messages.

"A. Carol 7 684-917. R."

"Raul, what does this mean?"

"Your friend, Carol, wanted to come over last night at seven. You were supposed to call her."

"Last night?"

"Yeah, you were in your cabin. And Luke needed the phone."

"I don't know a Carol. Pearl maybe?"

"Yeah, that's it!"

"This phone number only has six digits."

"The last one's a zero. I filled it in."

Using the phone after someone else gives clues about their conversation. Cuss words etched into the phone pad are a bad sign, as are nail marks on the cord. We turn up the music and pretend not to hear the occasional quarrel slipping in strained tones through cracks in the booth. When you live with others, you can't always hide your struggles and conflicts, and yet, you are not close enough friends to choose to share such things. So secrets are both shared and kept, breeding an unspoken security among us.

The lack of privacy is compensated by inaccessibility. Although I sometimes miss a coveted call, there's freedom in being hard to contact. I rarely have to talk to disgruntled or demanding people and have fewer social obligations. Phone and door-to-door solicitors haven't discovered us yet. During quiet evenings alone in my cabin, I can follow an idea through its forested path until it arrives at a clearing. I can read an entire book or write a poem.

Our isolation nourishes community among us. We tell our woes, successes, and puzzlements to each other. And often, just human presence is company enough.

It's not that we are removed from the world. We all go to town sometimes, some daily, and a few even travel. Luke gets the local paper delivered and Raul listens to news on the radio. We have political discussions and bemoan the world situation. But Siesta Lane buffers us. We invite the world in; it doesn't bang down our door. And when it begins to crush our souls, we can ask it to leave.

Mick, Raul, and I particularly like to sit quietly around the fire. We've come to appreciate each other's laconicism. Sometimes I knit, Raul cures leather, and Mick keys out plants. Other times, we just stare, in hypnotic spells, flames reflected in our pupils. A few of us may play cards together on an evening, or read short stories aloud. Meanwhile, the world's discouraging news curls and disappears in the woodstove.

Perspective

Natalie and I write up a treasure hunt of the land. At first we make clues only for familiar places on the land—"a high-up house" would be the treehouse, "water's world," the pump. Communication is often tailored to what others will recognize. This way there is less to explain. And yet, by narrowing our talk, we reinforce only the small known threads of the web. We see only what we talk about.

We decide to write clues like these:
1—the coolest spot on the land
2—a good sunrise-watching place
3—palace of the spider webs
4—home of the rare plant
5—tree with the most birds at dusk
6—the greenest softest patch of grass
7—tree with the water-filled hole squiggly with mosquito larvae (an attendant spider web poised above)

Such details of the natural landscape used to be invisible to me. But now the hidden world beckons, requiring

patience and a retraining of my eye, a refocusing. To see in the shadows, pull the periphery center stage.

We give out the treasure hunt at dinner with mixed response. Mick knows where the rare plant is, and Sara the tree with the most birds. Luke wants to put a fish in the mosquito breeding ground. Paul suggests a good sunrise-watching place, though it's not the one we had in mind. Raul, in his understated tone, suggests that the coolest spot on the land would be the freezer.

Fire

Lantern light softens a room, scents it, offers a flame to stare into and a place to warm your hands. Shadows smudge the walls.

Last night a forest fire pinched its way north over the ridge—a caterpillar of smoke with glowing belly. I sat helpless, two miles away, watching the helicopter's blinking lights hover pathetic in the smoke. What are we to fire?

Smokey the Bear with his accusing scowl, and wide-eyed, slender-legged Bambi, running in terror from the flames, have convinced us that fire is *bad*. Yet, historic fire often helped plants and animals that benefited from the resultant patchwork of clearings in the woods.

Just as we have failed at stopping floods, we haven't eliminated fire. Most ecosystems have evolved with fire and floods, and in a natural system such dynamic forces strengthen and animate the land. But those fires tended to be less extensive—burning snags, clearing brush, licking the base of fire-resistant old growth. Without the buildup of fuel from fire suppression, most of them had a lighter touch than today's infernos. They pruned and revitalized the forests.

Fires have sculpted the Pacific Northwest forests for millennia—some sparked by lightning, others by native people to increase deer forage and food plants, maybe even to allow a clearing for a sunny camp. Many species benefit after a fire. Fireweed and other pioneer plants thrive in new meadows, leaf litter is cleared away allowing for seed germination, and in some plants, like the ponderosa pine, fire stimulates their growth.

Oregonians hold a constant debate about wildfires. Let them burn? How much? Remove woody debris before fire season? Set controlled fires during the wet spring?

Today, the results of logging and fire suppression (i.e., lots of diseased and same-aged trees) cause fires that burn most everything in their acres-wide swath. Even these fires bring gifts to the forest of strong light and vibrant new growth. They are a drastic response to dire conditions. Yet we fight them as if they're demons from hell—blasting roads through to get fire fighting equipment in and dumping chemicals on the flames and soil, often to no avail.

And the increasing practice of building homes in the woods has firefighters risking their lives to save private property in situations where they would otherwise let the fire burn itself out.

Lightning storms. Floods. Tornadoes. Dramatic natural events thrill while they frighten, with nature pushing up her sleeves. They humble us who imagine we've got the cat in the bag.

When too strongly harnessed, our own flames— anger, desire—can rise, spring from their campfire, and rage around the room. And then they explode in actions we regret and didn't see coming.

So I try to lose my own fires, to let them out on a rein and wheel them in, but with a lighter touch—to mix anger with humor or take long brisk walks fueled by passion. In my single state, allowing myself to be aroused by life. To let the Eros of quivering leaf, soil scent, and lush moss fill my senses, ring my bell.

Our music and literature narrows desire into a sexual realm—heterosexual, specifically, between young adults. And we must be beautiful, so the movies suggest, for this to happen. And the beauty must be of a particular shape and design.

Sensuality tingles my skin around men yes, *yes*—but also when touching the veins on my mother's hands or hearing the cadence of a friend's voice. And when dusk sweeps red lashes across the sky and breeze sends fingers up my shirt, Eros plays her seductive music through this body's flute.

So wasn't Walt Whitman on to something? When we constrict desire to the "lover," focusing all loving attention on "him" or "her," trouble brews. Jealousy. Insecurity. Clinging and clutching. End of relationship.

I once told a lover that our touch gave expression to the emotion generated by what I saw and heard all day—the pearled dew on the snowberry leaf, fluted gills on a chanterelle mushroom, silhouette of a hawk against a cloud—that our intimacy was a place for these kindled feelings to flare and reside. That it went beyond what I felt about him.

"You mean you're thinking about other guys when you're with me?"

I never tried explaining this to a lover again.

What would happen if we allowed desire to ripple our emotions with a constant breeze? So that we lived in a state of desiring what we already have. All our friends and family and this eager lover—this pulsing, fragrant earth.

Perception

Trees do not go straight up and down. This is one of many illusions created when we look at nature in the media or out of car windows.

On my hill there's an oak that extends horizontal to the hill where it anchors. It grows exactly sideways, sticking out like a ship's prow, with a girth of at least a foot. I could climb it and stand watch over the field. I could do pull-ups. I stop thinking about what I could do on it, and begin to honor it. Oak of the Long Arm. Oak Pointing South. Oak Wizard. Overcoming Gravity.

Another has limbs twisting and flowering out from the trunk. This one I climb, scraping my knee on the way up, a reminder of the difference in our skins, I feel, the tree's way of showing its mettle, how tender are we humans. Once atop, I curve into a moss of legs and arms and body, wondering about this ancient form of yoga, this stretching and perching in trees. Our bodies were made to be cradled like this. See how weak our knees, how curved our spines.

An ecology teacher I know requires his students to sit in one place in nature for twenty-four hours watching, listening. They notice where wind sashays, where sun

shifts her shawl of light, the least sound magnifies. Day
ebbs and flows into night.

The Wider Lens

Although the way we study detaches
bud from twig, fossil from lake, still
I have not learned to separate
the dropped pine cone from its quilt
of fir and maple leaf, trout fin
from river rock, birdsong from
dawn light. Nor can rain smell
be severed from bare feet, snowfall
from red cheek, or the lake
from the wind combing its skin.
The arms of the manzanita cannot seem
to untangle from the horizon's grey waist.
Because the speckled gall changes once it slips
from its shelf of oak bark to rot
on mud ridge and twig scatter. And
the curled fawn isn't the same without
the calyx of grass against which it rests,
or the hemlock beyond that, any less the owl
in its top branch asleep with yellow eye open
on the mouse to be churned into a furred
pellet and spit out beneath
the dark cape and its
circling aperture
taking it all in.

I'm often late for appointments with friends in town
because everything seems slowed down on Siesta Lane.
The drone of crickets, waft of grass, the horizon's long
breath. As the days amble past the more I gaze on the
field, submerged in a reverie of beauty, knowing that

whatever needs doing can wait for this small homage. Duties here are more basic and labor intensive. Haul water. Split wood. Hang laundry. Life becomes not more simple, only more essential. And there are new responsibilities to claim: the spider webs to dodge, the furious bee to free from its battle with the window, the backrub for Sara who's been weaving too long, watering Raul's cilantro plant, petting the dog, Ceres.

Last week when I opened the washer, a dazed lizard stared up at me, its toes clamped to a wet towel, back arched and tensed. He must have been living in my laundry basket. Freeing him onto the grass, I marveled at his tenacity through the spin cycle—and puzzled over how to get the yellow lizard juice stains off my blouse.

Alliance

I hadn't wanted to keep the dog Ceres around. I'd wanted wild birds and no barking, to leave my cereal undisturbed on the deck. She came with the place, Quin said. So I ignored her for awhile, except to moan about her ferocious announcement of guests.

But lately she's been following me, her wet chestnut eyes in triangles of appeasement, ears akimbo, head atilt. Like a gallant gentleman, she escorts Sara and I, the only two single women, to and from our cabins every day. I begin to love her.

Off on a walk, her heels click behind me. I am thinking to myself, "No, Ceres, you stay here." I neither motion to her nor alter my step. Still, she stops immediately, and turns, returns to the house.

But how does Ceres understand if not by words or gesture? She seems to read my intent. As if she understands me at some pre-verbal level. I am left astonished and accompanied on a dusty road with myriad small bright eyes pointillating the woods around me, humbled and caught in a net of awareness of which I had been oblivious. I call Ceres back, and stroke her head and sides, surrendering another thread of affection.

Siesta Lane

This is spider territory—where invisible strands web me to what was once blur and background—the oak cradling my cabin, clouds on the hill, that sassy woodpecker in the pine. An orb-weave spider will draw steel-strong silk from her nozzle-like spinneret and weave complex webs daily in any suitable opening. Some strands will be sticky to catch insects, others smooth enough for her to walk along, others white as a warning to birds. She'll sense the suitability of a mate from the strength of one's pull on her web. If he's not careful, she'll eat him.

I have been walking more gently on vegetation, gingerly removing ants from my arm, leaving the impeccable deer antler where I found it in the woods. I am caught by love, here on this land where I've been living through so many quiet, dangerous hours, not knowing invisible nets were being cast, that the heart would have no refuge.

Immigrants

Autumn means apples. A tree a mile down the road on vacant property blushes with its heavy apron of fruit, tipping its crown as if to ask for help with its load. One late afternoon, I climb and linger and munch away the darkening day. Fall at dusk, spring at dawn, winter nights, summer days—double their potency, with autumn afternoons, for me, most poignant.

These apples are green and red, tart, crisp, bumpy, and small. They could be one of dozens of varieties or a hybrid. Pioneers brought apple starts on wagon trains, sheltering the seedlings against cold, watering them even when water was scant. Apple trees could provide fruit, vinegar, shade, wood, blossoms, and alcohol to families for generations.

These introduced plants did no harm to native species and quickly became part of the lore and habitat of the West; unlike *Homo sapiens* who arrived in North America over twelve thousand years ago. We spread across the continent and encountered native species, such as mammoths, mastodons, horses, camels, ground sloths, lions, giant wolves, great bears, and saber-toothed cats, that had not developed defenses against humans and slaughtered them in large numbers, depleting our own food source and messing with the ecological balance.

Native American stories often stress moderation, a light touch, humility and gratitude toward the natural world. Such an extinction would have been an argument for a shift in paradigm, for a value system based on balance and sustainability. Even though they too have altered the land, North American tribal people seemed to have achieved a basic harmony with this continent before they were almost killed off by the wave of Europeans that washed across it starting in the fifteenth century.

Had the European settlers heeded this hard-earned wisdom and not massacred the buffalo, plowed the prairies, drained the wetlands, poisoned the air and water. We, like the apple tree, might have been a welcome new member of the community with whites and natives protecting each other, and the land, for the future—some of which is here now. Our current society might have looked something like Siesta Lane.

Twilight shoos away dusk with her shadowy hands. Laden with booty, woozy with apple dreams and fading vision, I pick my way in the growing dark back to my borrowed home.

October Crickets click outside, and inside, burnt wood drops to the flames. The fire burns what has been blocking me—fear and need, the desire for freedom and affirmation. Outside in the dark, juncos flit from branch to deck, clutching fat worms in their beaks. One eye tilts to me as if to say, "You'll be one of us soon," with the forest settling in the rain, the worm swallowed.

Earth

Chemical and biological weathering produce yellow/red iron minerals, black manganese/sulfur/nitrogen deposits, or brown organic compounds. Loam is a fertile mix of clay, silt, sand, and organic matter; Loess—a fine-grained yellowish brown deposit of soil left by wind; Till—sediment deposited by ice . . .

I am developing a dirty mind.

Becoming more and more aware that everything, really, depends on the quality of the soil. Whether or not our great-grandchildren will have great-grandchildren, the state of our internal organs, art, government, romance. If unassuming lichen doesn't chaw granite into accessible bits and microorganisms don't fortify these with dead organic matter, plants will disappear. And without plants, no humans.

Aware of this in my bones, I dig potatoes today, like an Easter egg hunt, flailing dirt and shuffling for pale red globes in the moist soil, some tiny as pearls, some big as fists. What happens to make one potato boast a swell in the earth, while its neighbors pinch and pucker in? Yet the big ones are game for moles and groundhogs and the wary stab of the pitchfork. Those near the surface have turned green in the sun, gathered like billiard balls in their crumbly pockets.

Serious business, potato digging. Each spud enough, in a pinch, to nourish someone for a day. And for that we planted them, and cheered over their ample green laps spreading in the August sun. Because potatoes smell of earth, fit in your palm, fill a tummy. Once harvest starts, I want only them, raw as possible, with chunks of rock salt and crushed pepper. I crush the pepper in a mortar just enough to open the corn, fill a pot with water and potatoes, and boil them until the kitchen grows moist with potato vapor. And as I lift the lid, steam makes my face rosy and slippery as the softening spuds. I dip a whole tender potato into the salt, into the pepper, and into my mouth. I eat with my fingers, still with garden dirt in the creases, bent over the table, my whole body a chasm that can only be filled with potatoes, potatoes, salted and peppered and steamy, skins sliding off the opal flesh. Potatoes just up from the garden which has collected my sweat and song, where I have hoed through crisis and calm, whispering to the plant which has yielded me its root, to eat, to give us this day, amen.

Symbiosis

Digging around in a nearby woods I find a cluster of wizened Oregon white truffles the size of pool balls. They smell like must, like walnuts, and my imagination adds garlic, onion, and olive oil. I take it home to check it, again and again, against the pictures in my mushroom field guide before convincing myself (but no one else) to eat them.

These potato-like fungi grow around the roots of the oak, enhancing the tree's absorption of water and nutrients from the soil. In turn, truffles borrow sugars from the tree, and perhaps take advantage of acorn-scavenging squirrels for spore transport. Truffles, truly welcome guests, are the prime example of one kind of symbiosis: mutualism.

Symbiosis occurs when species live in association: one harming the other (parasitism), one benefiting without affecting the other (commensalism), or both benefiting (mutualism).

Oaks have all three forms of symbiosis. Mistletoe, romantic to us, is deadly to oaks. It wraps its lethal arms around the limbs and suffocates the tree. Mistletoe seeds are disseminated by birds that wipe their sticky, seed-filled beaks on limbs after gorging themselves on the red berries. Parasites, truly.

Opportunistic but harmless gall wasps demonstrate commensalism. They scratch an oak twig, lay their eggs, and leave a secretion that triggers a chemical response prompting an oak to grow a round scab, like a brown, freckled ping-pong ball, around the wasp eggs. This scab, or gall, provides protection for the larvae until they peck their way out as adults in spring, with no harm to the tree.

To foster a form of mutualistic symbiosis at Siesta Lane we hold meetings to split chores and bills and to check in. But we find that our heritage of rugged individualism makes it way easier to nurture a particular peeve than to think about group stability.

"The bathroom smells moldy."

"SOMEBODY'S WEARING MUDDY BOOTS ON THE CARPET!"

"The last person to split wood left the axe out in the rain."

"The oven cleaner's shit." "It's environmentally sound." "It's still shit."

"How come the water bill's so high, Bath Queen?"

"Could someone besides Raul take out the garbage for once?"

"I forgot what my chore was this week." "Check the list!" "I forgot where the list was."

Knowing parasitism will destroy us, we try the truffle way.

Sara is sick for a week so I bring her soup and clean her cabin. She knits me a scarf.

Raul's truck breaks down and Luke drives him to school. Raul teaches Luke's class so he can go to Los Angeles for his folks' anniversary.

Siesta Lane

Jack uses more wood so he chops more. Mick puts bouquets on the table in return for our help with plant collecting.

The communal spirit is a spider running on slender legs across the threads of the web, strengthening here, slackening there, making sure the relationships can carry her shimmering, necessary weight.

Choices

Going into town after a week in the country makes me dizzy. I have to steel myself against the traffic and pace of what is not a very large city, Eugene.

Shopping is particularly agitating. Why must there be thirty kinds of cereal? And how do I weigh the merits between low-fat yogurt with acidophilus culture and honey-sweetened nonfat yogurt, or a lotion with aloe vera as opposed to one with coconut oil?

And afterwards there's the second guessing: "Maybe the rosemary-garlic vinaigrette would have been a more versatile choice than the sesame shiitake ..."

I had always believed that the more choices in life, the better, but now it doesn't seem true. It feels more like oppression of its own kind. Every minuscule decision takes time and energy, takes me that much farther away from my writing, the land, the people I love, and my connection with everything deeper in life.

My Norwegian friend, Pär, who may be the reincarnation of John Muir, explodes every time he eats in a restaurant in our country. He opens the menu and starts pulling at clumps of very blond hair and twittering his leg. "Just bring me some toast," he pleads with the waitperson, quickly folding up the extensive menu. "Would that be white, wheat, rye, sourdough, or pumpernickel?"

The vague sensation haunts me that I once reveled in the variety of choices open to me. What people to befriend, jobs to take, groups to join, places to live. Looking back on my various relationships, moves, and jobs, such options appear to have cultivated breadth at the expense of depth.

Before moving to Siesta Lane, I lived in ten different places in the same small city. One day a guy looking like Rip Van Winkle showed up at my door with the mail.

"Hello. This was sent to your old address."

I looked down at the address, which was at least three moves ago. "How did you trace me here?"

"Well, I been a mail carrier for awhile in this town, yours a lot of the time. Name's Ned. First delivered mail to you at 1020 Lincoln Street—nice place! Then 804 Chestnut Street, then you moved in with that older lady at 471 Atkinson Avenue, lessee, then down the street to the upstairs apartment at 1219 Beech Boulevard."

I shrunk backward into the latest of my hermit crab homes. Ned tucked his very long beard into his postal shirt and adjusted his bag. "Guess I got a mind for numbers, doin' this job. Nice day, Ma'am."

I'd been found out.

Lost in a maze of possibilities, for years I couldn't commit to any one person, place, or lifestyle. Now it seems less and less like a game or a freedom to have so many choices, and more and more like a carnival ride that won't let you off.

Rain

Rain has many sounds. A low hum as it strums the landscape. Loud plips where it collects on the corners of the eaves and falls, and where the ceiling drip meets the pot. Then the crescendos of windy rain, the chatter of it on the roof. A stormy opera.

I walk down Siesta Lane and the rain flows over and down my body, hurrying into the least crevice in my clothes. What else can I do but weep with the current? Sorrows I don't remember drivel out of me, course down my face with the rain until I am washed clean, inside and out.

Rain taps the roof, pads the path, knocks on the branch. I am not alone in rain. It drums my head, strokes my face, beckons, cajoles. It is what in nature reaches out and touches. It is the voice of the sky. I can cower and shield myself, dodging into shelters, or I can surrender, and send my soul to greet it.

The natural elements beckon and bless what we hide, wanting to know all of us. We can be carried in their stream. Bird. Cloud. Wind. Rain.

Or we can run to the city where scenes buzz with action, where distractions whittle away at our emotions.

Amy Minato

But here feelings grow wide, roam the meadow, wisp into the woods; drawn out by the quiet horizon, they can stretch to the sky's end. Sorrow spreads and mellows. We steep in it, the ache in the marrow, tears in the rain. Joy flies in the quiver of leaves on an oak, bounds as a squirrel to a branch, catapults from the white mouth of a cloud spewing sun.

To live here means having the courage to be aware of your heartbeat, terrified at your mortality, each sense and emotion tuned and strummed.

No wonder some turn to drugs, television, shopping. How impossible to sit in all awareness in our delicate skin, pelted by the incessant rain.

Sustainability

Sara comes home from a talk by Helen Caldicott about the ecological crisis and is ready to change her life.

"We can never use plastic again."

"Okay," I say. This is right up my alley.

Now, I've heard of an eco-guru who had everyone in his intentional community scared to use toilet paper. He told them that they should be able to wipe themselves sufficiently with one square of TP. Guilt compelled them to do it, and to use only one corner of a towel to dry themselves off, and to not eat anything that needed refrigeration. This was one guilt power trip, but it worked on some people, because we have this hidden wound over how we've impacted the earth. For some of us the wound is less hidden, for some it's buried deeper than a kudzu root. Sara and I were interested in realistic changes, simplifying one's life as a challenge not as a burden. A gift that ends up, in a crazy way, being for yourself.

Together we eco-evaluate our home. First, we elaborate on our recycling efforts, setting up more specific receptacles: clear glass / colored glass / plastic recyclables / non-recyclable plastic / metals / glass containers with

84

deposit / metal containers with deposit / containers to reuse / low-grade paper / high-grade paper / newspaper / milk cartons / real garbage.

Then, we make sure all our food waste gets composted. (Only Raul, who claims it as part of his heritage, eats much meat.) We sew small bags using fabric that has bright colorful fruits and vegetables on it for buying bulk foods, and bigger, heavier ones for carrying groceries home. (Our TV-free home is conducive to fun craft projects.) Risking pneumonia, and having few visitors, we keep our cabins cool to save wood. Finally, we beseech the others to carpool.

But that is where our eco-community unravels. In our household, five of the eight of us commute twenty miles per day to work or school, each in separate cars. Sara and I plead, we set up schedules, suggest specific carpools, but everyone's needs are different, there is no city bus to Rainbow Valley, and even veteran bicyclists would have trouble with the steep hill riding home. Sara and I feel deflated and implicated.

It seems that we, as a society, have grown accustomed to convenience. In fact, we consider convenience an inalienable right. And we've made our lives crowded enough so that we can't meet our responsibilities without it. Also, we love to move. "Always on the go."

All this adds up to a dependence on one of the most addictive and dangerous, albeit wondrous, inventions of our century: the personal car. We are wearing a strait-jacket, which, as we twist and turn to get out of it, wraps us tighter.

Living a simple, ecological lifestyle requires minimal use of a car. My next home, if I'm savvy and lucky enough, will be within walking or biking distance of most of my needs.

But for this brief, blest respite from soul-crushing capitulations, I stay home when possible, and listen to the land.

Limits

Ten hangers will hold four dresses and six blouses. Pants must fit in one drawer, shirts in another, sweaters in the wicker basket, books on the shelf.

Rather than expanding my storage space to hold my stuff, I've put a harness on my possessions. Everything I own needs to fit in the confines of my cabin. No more hangers, no more drawers.

In a country where the average person acquires about fifty-two clothing items per year, this is a radical activity, oppressive at times. But the result is sanity, clarity, peace. For each item discarded, a sign reading "Liberty" swings gaily in the wind.

Any books I am unlikely to open get donated to the library. When I buy a new piece of clothing, another is given away. Changing possessions keeps my creativity alive. Getting rid of as many as I acquire sharpens my will, and my practice of nonattachment. I now have a Zen wardrobe.

Sometimes I'll miss a favorite shirt or dress, and wish I'd kept it. A portion of my attire is infused with meaning—because of whom or where they're from, where they've been and with whom. My multi-patched jeans from high-school wilderness trips, the red Guatemalan shirt I wore attending a birth, the indigo vest into which I knit all my passion for an unavailable blue-eyed man— desire stitched into its soft weave. The ones made by hand or from distant lands. These I try to keep. To mend and mull over. When I don such clothes I am wearing memories, reminding myself who I am and have been.

And so I forge a tentative harmony with possessions, floating together like debris in a swift current. All of us flowing downstream. Scattered to the ocean in the end.

Sharing

Inevitably it will happen. You'll be making dinner and need another egg for your cake or a teaspoon more of lemon juice for a sauce. Your guests are coming over in five minutes. No one's home to ask, so you borrow the egg, the lemon. You reach onto a shelf marked with someone else's name. You'll square with them later.

We learn a lot about each other when this happens. Whom you can never borrow from, whom you can always borrow from, from whom you can only borrow certain things that are impossible to discern until after you borrow the wrong thing and get scolded. Who needs you to replace stuff now.

We learn who would freeze through dinner before borrowing a sweater, who would rather borrow than use their own things, who borrows without mentioning it, who can't remember what's theirs, anyway. Who's not careful with stuff. Who is.

A salad spinner gets borrowed and broken and all hell breaks loose. The owner is steamed and everyone has opinions. The tension around dinner prompts a discussion of property values and we draw on our childhoods for explanation of our differences.

"We were punished if we broke anything."

"Everything was common property in our family."

"My sister and I fought over possessions."

"It was okay to borrow something but if you wrecked it you were screwed."

"All our stuff was junky so it didn't matter."

"It was honorable to share."

"It was rude to borrow."

The hard feelings around our different styles begin to soften. We learn about each other.

Sara will put away her treasured items.

Be careful with whatever you borrow from Luke.

Amy wishes that most things were shared.

Replace soon what you borrow from Paul and Natalie. They have plans for it.

Raul doesn't believe in private property.

Never borrow from Jack and Rita.

So rather than a set policy we juggle our differences. Tough to keep straight but a way to respect the individual within the community, and our varying degrees of need for self-sufficiency.

Meanwhile the natural world waits to be included in the exchange—bees pollinate our vegetables, mosquitoes draw our blood, and wolves recede from our roads. No requests, no explanations. But wait. Was that the squirrel calling for a house meeting?

Food

Because we are a group of fair-weather farmers, not all the vegetables we eat leap onto our plates from our garden. Also, our garden water is somewhat saline which inhibits quite a few of our crops. But that's an excuse. Mostly we Siestans are half-hearted gardeners, leaving the serious row-hoeing to experts.

Siesta Lane

In Oregon and other states you can become a member of a farm. Community Supported Agriculture is a system where a group of families and individuals buy shares of a farm's yearly crop. You pay up front for a season's worth of fresh, organic, locally grown food.

From May through November you pick up your delectable variety of fruits and vegetables from a designated spot in your neighborhood (usually another member's porch or garage). You trust luck about what and how much you'll get each week. Exotic new vegetables, exciting recipes, and sometimes phone numbers are passed around. Everyone's invited to visit or help out at the farm and to come to the few festive gatherings each season.

One day I visit the couple that runs Full Circle Farm. Jude, with a seven-month-pregnant belly carries equally rotund pumpkins under each arm, stacking them onto an old truck. "Usually, we deliver by bike trailer," she says, "but not with pumpkins!" Her cheeks flush, her eyes shine. Martin greets me, grinning through his mop of hair and thick beard, wiping cracked hands on worn overalls.

I join them in a potato dig. Jude shovels up the soil and we scramble for spuds, brushing clumped mud off them with our gloves, filling box after box.

Large, dark clouds bunch and bustle above our small crew in an autumn field. Within the flat expanse of land and sky, our three hunched figures seem inconsequential, whispering and gleaning slowly along the rows. The rain starts and we head to a shelter for lunch.

Jude has cooked stir-fry on a propane burner. They lease the land so there's no house here. Volunteers have helped rig up this rustic shelter.

After lunch we sort potatoes beneath it, pitching them into piles by size. Martin separates the purple ones for seed stock. Jude fills the bags. We laugh and swap stories. The rain ends. The day passes too fast.

They send me home on my bike with pockets full of garlic and kale, good conversation, and deep respect for their work. Independent, organic farming is a rich and difficult life. Sharing it for just one day has given me a new measure for my own.

Happily, the dirt in my palm creases takes days to fade.

November The western screech owl's call is more a stutter than a hoot, a repeating monotone "who-who-ho-ho-o-o" that picks up speed but loses volume, like a bouncing ball. It punctuates the quiet night, coming from the soft dark of the trees. I feel compelled to respond, to answer this tiny feathered creature's haunting insistence, but have yet to discern its offering, to begin to comprehend it.

Snails

Each morning snail trails decorate the entrance to the main house, curling around the welcome mat. The sun lights these silver hieroglyphics, messages from their nocturnal wanderings.

Why do the snails come *here*, I wonder? Do they want in? Are they trying to tell us something? Or does the cement stoop provide the perfect dancing pad for courtship? Snails are hermaphrodites, so whoever shows up is fair game.

I like to imagine that they mate all night outside our door. And with snails, it may take all night—upright with their gooey fronts sliming together. For whatever reason they come, I look forward to their secret scribbles brightening our doorstep.

Snails are honored in Buddhism, as they climbed onto the Buddha's head while he meditated to protect it from the sun.

Siesta Lane

Buddha statues show snails arrayed like curls, spiraling around his crown. There they lay with their hard shells and soft insides. What better emissaries of stillness? Creatures that take their time have my recent admiration. Enamored, as I am these days, of slowing down.

During several dynasties in ancient China, you could tell the wealthy because they wore long beads around their waist. This kept them in dignified motion. If the beads clanked together, they were walking too fast. Being rich, of course, meant you didn't have to hurry.

How opposite it is in our country today. Being successful means you can answer your phone fast—it's on your hip! You can move across the world faster than birds, you can send e-mails instead of waiting for the postal service.

In the city park I watch a determined mom run along a circular path bouncing a baby in a jogging stroller. The wide-eyed child looks as scared as my cat does when I have to drive him somewhere. Quixote's tongue hangs out and he leaps from window to window before burying himself under a seat. He knows we mammals aren't designed to move so fast, except for maybe a few of his cousins, like the cheetah.

When cars were first invented, people scoffed. Why would anyone need to go anywhere as fast as fifteen miles per hour? "The faster we go, the less time we have," a character bemoans in *The Magnificent Ambersons*, a classic Orson Welles movie. And there is evidence that those of us who regularly use computers grow impatient with the rest of life happening in real time.

A few years ago in Mexico a friend and I walked down to her village to buy the day's food. We strolled through cobbled streets, stopping at the coffee vendor, the fruit stand, the post office, and finally the tortilla shop. In each place Alejandra knew the owner, and would visit several lively minutes with them, catching up on family, friends, and local politics before moving on.

In this way the townspeople "polished each other's hearts," I was told. People of all ages and aspects charmed and flirted with each other, keeping their community healthy, their members feeling valued and looked after.

Robert Pinsky argues that "America is the idea of motion." We hit the ground running back at Plymouth Rock and haven't slowed down since. We're a restless tribe, looking for Eden across the next hill, destroying the land where we are because we're on our way somewhere else, anyway.

I once rode in a speedboat on its new run into China from Hong Kong. The dignified figures in wide brimmed hats beside their water buffalo may as well have been from another century. They looked up and stood completely still, up to their knees in mud, as we raced past their rice paddies. Most likely their family had been growing rice on that spot for centuries in the same careful manner. Our boat spewed oil into the river as we stared back, with the incomprehensible luxury of tourists, feeling, in my case at least, more than a little silly. "Death will catch up with you anyway," I imagined them saying. Who did we think we were fooling?

At Siesta Lane today the rain is a lullaby. Catkins hang and sway like gold lanterns on the hazelnut tree's filigree of branches. Rivulets circulate the meadow as my breath livens the blood in each capillary.

These meandering loops of water, blood, slime, the graceful trails I follow.

Cravings

"Are you thinkin' what I'm thinkin'?"

"Chocolate cake."

"Yep."

"Brownies. Fudge sundaes. Cinnamon rolls."

"What's in the pantry?"

"Rice cakes. Stale graham crackers."

"Oh."

"Any cocoa?"

Pretty soon Sara and I have created "Graham Éclairs." Graham crackers with chocolate sauce sprinkled on top and peanuts. This temporarily assuages our sweet teeth. Our craving for sugar makes sense evolutionarily. It kept our ancestors from eating unripe fruit and dispersing immature seeds. Today, it fuels our economy.

Being out of town forces us to let the wave of our small addictions and indulgences pass over our heads like a swarm of gnats, or to improvise.

You get through one day then the next without TV, electric heat, ice cream, or whatever you're craving, and eventually you hardly think about it.

I invite a good friend to visit me one weekend and add, "Can you believe I haven't had chocolate for a month?"

Amy Minato

"Maybe I shouldn't come," my friend replies soberly. I wonder if it is the lack of chocolate, or the thought of me being without chocolate that makes her uneasy. Finally, deducing my veiled SOS, she brightens. "You want milk or bittersweet?"

Although I consider my cravings fairly harmless, it's nice to tie them up in the stable when necessity requires, and a joyride to let them out after.

Survival

Oregon has the ninth-highest number of endangered species in our country. A few of these plants: Bradshaw's lomatium, Willamette daisy, and Kincaid's lupine live in the few remaining wetlands, one of which lies just over nearby Gimpel Hill, an area that the Nature Conservancy has recently burned in an attempt to help reestablish native plants. Use of fire was a common practice of the early native people of the area, the Kalapuya Indians. In this way they would keep back some of the trees encroaching on their, and the deer's, food plants. The Nature Conservancy mimics this technique to assist its conservation efforts.

This month, the prescribed fire kills several ash trees that need removal. The Siesta coalition plans to help open the land by cutting down a few of the dead ash trees for firewood. Mick, our lanky live-in botanist; Luke, the printmaker from Los Angeles; a rented chainsaw looking like a large piranha on the truck seat; and me, whose closest experience with heavy machinery was driving a Zamboni ten years ago. We make a fearful foursome.

We begin at 10 a.m.

Amy Minato

By 10:30 a.m. we have gotten Luke's vehicle determinedly stuck in a gravel ditch. By noon, and after sufficient awkward maneuvers, the truck is free. Undaunted, Luke and I clamber into the back and bounce along over the potholes to stake out a few slim, straight trees near the road.

The trees stand sentinel amid the brush, pointing charred beaks to the clouds. We find where the creek has been, and look for the beaver that is hidden away, her home dried up from the long summer. We pick a few splotchy apples from a wild tree, ones it seemed unlikely other creatures would harvest. They are pinched and tart, keeping their juices to themselves, their sweetness in reserve. As people do, who are unloved for too long.

We rev up the chainsaw and scatter the guardian jays that chatter furiously back from their new perches higher up. Luke, with blind courage and considerable aplomb, puts on goggles and earphones and hunkers down against a tree base in a cloud of engine roars and wood chips. Samaras, ash seeds shaped like boomerangs, waft onto his shoulders. Drops of thick red oil spill onto his jeans. After an alarmingly short time, the first ash tilts and thuds to the ground, where the grass beneath it seems to arch up and cradle its fall. Luke turns off the motor. The birds quiet.

We stand in the meadow's peculiar silence, looking at what two minutes had dropped, which had been growing, until the fire, all its years toward the sky. I sense we have somehow startled the landscape, that it shivers and shifts slightly, allowing room in its consciousness for this new intrusion, this necessary surgery.

Luke restarts the saw and sweeps it like a swashbuckler from base to tip of the horizontal tree, rowing through the largest limbs. Mick and I snip the smaller branches off the trunk. No one speaks, but a humility is born among us, by this act bound to our surroundings.

Couched in our tender flesh, we are three beings helpless to the elements without clothes, shelter, fuel. We are grateful for the chance to acquire firewood in a sustainable way but we are using the remains of what had once been alive on earth for the energy to do this. Our faces alternately flush from sweat and cold, we remember the winter coming on. I take my turn with the saw.

Compensation

I'm donning the requisite layers of clothing beneath my extensive raingear. My mitts are even waterproof, and one could store cider in my industrial boots. It's the Pacific Northwest, after all, and rain's the lover who won't stop tickling you.

On my walk I pass ducks showering blissfully on the pond. Every once in awhile, one freshens up its feathers with water repellent from the oil glands on its back. Why don't we have such a neat trick? Fur to keep us warm or large teeth to frighten predators? Nobody else in the animal kingdom wears clothes.

I think of the exquisite swirl of pinecone bracts, their shape adapted to coax windblown pollen to the ovaries. Or the graceful guise of the owl, whose soft feathers make no swish to alarm its prey. There's the rough, water-repellent fur of the wolf, the whale's inspiring blubber, the curved claws of the coyote. But what about humans? Standing upright was a good trick, as were our opposable thumbs. But in some ways, haven't we become less adapted to the natural environment? Our vision dims, hearing is less acute. With a comfort range of less than

10 degrees (sans clothes), we must be more sensitive to temperature than our hardier ancestors.

That's the niche we fill, I guess. The oddball cousins with overextended brains to make up for vulnerable bodies. Instead of relying so heavily on physical adaptations, we change our environment to adapt to us. We create things to protect us. Clothes, fuel, weapons. Our bodies lose resilience.

Other life-forms evolve, adapt, and equip themselves from internal clues, without major tinkering with their environment. Seen this way, the evolution of other life on the planet can seem more profound and accurate, their adaptations simply *work*, without the residue of new problems. Plants and animals don't depend so heavily on analysis. They produce a genetic variation that works, or they go extinct.

For us it appears to primarily be the mind that evolves, and we then cater our surroundings to our weakening bodies. Wild nature feels alien because we can't survive in it without accoutrements from our society. For many people, camping now means watching TV in an RV in a state park lot. We manicure our yards and design golf courses, making nature palatable for us.

But many non-human life forms can't adapt so readily to our manipulations of the earth. They retreat to the rare undisturbed places or disappear from the gene pool. We stare sorrowfully at other species across the fence, the same barrier that gives us Mozart, Michelangelo, Shakespeare, but also pesticides, clear-cuts, and oil spills. We like our inventions, and we never meant to unravel the garden, but we are using its threads to build our cities and fuel our machines.

But can this go on much longer? There's only so much sand in the sandbox. Nowadays I gaze on wildlife with apology and a bit of envy. They suffer more now, but

when society bottoms out, who will be better equipped
to survive?

The rain has become a silver screen so I hurry home.
The radio reports severe flooding and landslides in the
Willamette Valley, exacerbated by erosion from logging
and developed floodplains.

I place my wet raingear and damp clothes by the fire
in the main house and note that rain has soaked through
to my skin in many places, anyway. No one's around so
I run nearly naked across the grass toward my cabin,
stopping just once to lift my face and surrender to the
storm, before a sword flash of lightning has me scurrying
off again.

Dread

Everyone knows there's a ghost at Siesta Lane. Shutters left open, spilled milk, laundry gone from the line, a distant, bitter laugh.

You may be alone in your cabin, listening to the rain, arranging books on a shelf, and she shows up—cold and pale, dripping on your welcome mat. Your insides liquefy like a spider's victim. She is silent, still, sullen, tenacious.

All night you are a stone dropped into a deep well. You moan and writhe and do not sleep. She is greeting every anxiety of your life at the door and beckoning them in. It is her party, at your house.

There's no easy out at Siesta Lane. It's dark and quiet as a grave for miles around. Fears sidle up and jump on you like lap cats. Wrapped in goose bumps, percolating with nerves, you pet them. Loss, alienation, mortality. They purr and snuggle down.

During an excruciating self-imposed exile teaching in Greece right out of college, I resorted to staring in the mirror reminding myself that I was going to die, eventually. I thought that summoning fear would keep it from trailing me, like staring down a bully or a barking dog. But instead it was the Medusa. The more I looked at death, the more I turned to stone.

So on a rare night on our homestead when the fire has died and everyone else is asleep, dread might pass through one of us like a river, clattering our bones. We don't speak of our ghost, but at breakfast it's clear who's been visited. The one munching toast slowly in the corner, there since dawn, cutting out ads for high-rise apartments in town.

Then chatter fills the kitchen and the sun lays warm palms on our cold hands. Only the fiercest angst can linger past noon and withstand the sparrow's song or the smell of fresh bread from a warming kitchen.

Beauty

Oregon's Willamette Valley can be bleak in winter—blurred and monotone—like a sketch that fell into a puddle. Edges are smudged, shadows erased, colors muted. The rain can be so oppressively constant that the idea of it being used for torture makes teeth-grinding sense.

I catch myself wearing grey sweats and grey sweatshirts every day for an entire week. This is easy to justify, as I've been nowhere near town. However, it's not a good sign. I've grown accustomed to the exquisite landscape bolstering my daily reverie. With the scenery washed out, I feel myself sink. Since moving to the rainswept northwest, I've developed a mild case of seasonal affective disorder, a winter lethargy brought on by the lack of sun.

To avert a downward spiral, I need to muster my creative forces and push back against the gloom. This morning I rise determined to seed grace and color throughout my day. A creative rebellion in this weather's prison.

First, I don my Guatemalan dress with purple tights and an indigo scarf, although I have no plans to go anywhere today.

Amy Minato

I wash and braid my straight hair, and secure it with a brightly beaded barrette. (Today, I will say things like "brightly beaded barrette.")

"¡Hola! Como estas, muchachos? Día hermosa, no?"

Luke and Raul are hunched over their cereal, fingers melded to their coffee mugs. Raul does not look amused. Luke grins to hear even my sorry Spanish in the house.

"Está un día caca y tu sabes," he says. (It's a shitty day and you know it.)

I pin on each of them a button from my stash. Luke gets "Subvert the Dominant Paradigm" and Raul "Creatively Disturbed." The opera tape that I play loud enough to jingle their orange juice glasses drives them away.

"See you, Amy!"

"Ciao, Loca!"

Maybe this needs to be done with more subtlety.

On my morning walk I gather wheat stalks, dried grasses, and cattail that I arrange carefully in a tall vase and place in a corner. I wash the living room windows, do yoga, and sew bright napkins for each of us from my fabric scrap box.

I knead both a loaf of bread and a poem for the rest of the morning, make a fragrant soup for lunch, and replenish the birdfeeder.

Stacks of freelance work mutter and pimple with impatience on my desk. Too bad! I am taking an aesthetic day.

Jazzed by poetry, with bread and curry soup scenting the kitchen, and Mahler's Third Symphony secretly unlatching the clean windows, I slowly recover from the iron grip of ceaseless drizzle. The rain is a prince in a silver suit. I put on a bright yellow slicker to dance with him, and my heart lays petals of whimsical joy on Beauty's altar.

December

The pheasants come out of the grass like puffs of smoke: one, two, three. They waddle across the road and disappear into the hedge. I look more carefully into the ditch backed by barbed wire than I have ever looked into bushes before. The grey and brown and green take shape into rock, branch, leaf. No pheasants. Like my thoughts today the pheasants enter the brush: solid, quick, lost once they cross my consciousness.

Motion

A maple leaf, yellow as a duckbill, twirls a quiet ten yards down and lands at my feet like an offering. There, there is my life, I think. May I sway as I go.

The cliff bears rivulet marks where trickles have etched the sandstone. Water gnaws and eats. What is large is not necessarily safe. What is ancient, erodes. But what is motion and fluid also changes. Heat slurps the waves. Wind cools the heat. Taking and taken from—all a wild dodge, a continuous theft. Sweat rockets off my skin, exhausted blood cells race to my lungs for oxygen, which I in turn pull from the leaves.

Eastern philosophy and Western physics concur that the essence of life is motion and rela-tion, that nothing is isolate or stagnant. Just as our skin sloughs cells at the rate of 100,000 a minute, everything is in the process

of maintenance, renewal, and transcendence—keeping balance, making new cells, and creating new forms. Life is a creative process—a dance between active agents, a co-evolution, a music.

At a planning meeting for a new nature education group, in the middle of a friend's crowded couch sloped to the center, I meet a dark-eyed ponytailed man with glasses held together by safety pins who tries not to crush me. A physics grad from MIT, Joseph, teaches me about motion—the pull of gravity and magnetism, the Coriolis effect, centrifugal force, inertia, friction. Unknown to me, this begins our own dance that will ebb and flow through the next two seasons. For now, I'm courting my new life.

On a trip to the Columbia Gorge, I follow a single drop of water from the lip of a waterfall to the base of its crystal tongue, centering my eyes on its clear swoop beside the cascade, a prismed tear holding together in a dizzying free fall. We travel, as Frost's poem says, To Earthward, from birth to death heading for that long stay underground. But not willingly.

We build skyscrapers and airplanes, climb trees, lose weight, grow tall to beat gravity. We bend our necks in impossible ways to catch sight of our bright feathered cousins gliding overhead. What views they have! But would we really see better from up high? Or would we just see more, but with less accuracy?

I'm wondering if the constant itch to be successful (i.e., climb the ladder) stems from a rebellion of our weight, trying to escape the body's heft.

But here at Siesta Lane gravity wins. We sink and settle like apples into the earth, dirt in our skin creases, stains on our pants, leaning in as the planet whirls and hurtles through space.

And there's magnetism here, a force of attraction or repulsion between objects. Paul moves toward me who makes him laugh. He moves away from my desire to have children. I move toward Paul who plants peas and builds fences and knows how to raise a daughter. I move away from his smoking and muttering about women. His wife left him with a child he hadn't planned on but who commands his deepest commitment. Attraction, repulsion. The switch of these forces within him has set his jaw, lined his face, grayed his beard.

His daughter, Natalie, likes me but doesn't want her father to. She hurls the concentrated power of her bright twelve-year-old personality between us. Although drawn to their closeness and charm, their centrifugal spin pushes the rest of us to the edge.

When an object is moving in one direction, Joseph says, its tendency is to keep moving in that direction. Hence we lean to the opposite side when a car makes a fast turn. Inertia.

So that's why we stay in sorry relationships, I muse, thinking of my recent past, my bruises from smashing up against men who were decidedly taking a sharp turn. As there's nothing rational about staying with a bad habit, it's comforting to know there are physical forces at work.

In relationships we are like so many molecules, with electrons to give or take, weak areas to bind with stronger ones. Bumping up our energy level with our personal gains or losses. We move toward and away from each other, clustering into something either life-giving, neutral, or cancerous.

But what currently affects my motion most these days, what slows me, is the brush of grass and wind here, the catch of color and scent, the sensory details of this heady place that arrest my mind and stay my wandering soul. The varied and complex friction of the land. I'm hoping to stay still long enough to feel myself move.

Ballet of a Life Cycle

Beyond the park where two kids spar
with yellow bats, among the reeds
and willows, heralded by the hinged call
of the red-wing blackbird, the female
red skimmer dragonfly rides
the underside of the male as they drop and pause,
in each touch thirty eggs born to water, a slim,
crimson finger tapping the surface of the creek.

In two weeks, the bristled
lower lips of the hatched nymphs
will snatch for mayflies, minnows,
tree-frog tadpoles in languorous paddle
among the rushes. They'll ricochet
the creek bottom propelled by water
shot from the ends of their abdomens,
crawl to shore, and split
from their casings like trumpets. The pale
new dragonflies will buckle onto leaf
blades, fanning wings dry and
brittle as mica. Unwitting insects
will fly into the basket of their hooped legs.
In time, each male will curl
the end of his abdomen under and drop
a jewel of sperm just below his heart,
and like a red compass seek out
and saddle a female, eight wings beating,
and after twining in delicate hover
above the lilypads and fat eyes of bullfrogs,
arcing the fine tip of her ovipositor up,
she will accept the gift.

Homemade

"You're gonna make your own *mustard?*" My friend Pearl asks, placing a bouquet of irises on the kitchen table. Pearl is someone who quietly makes every place she visits more beautiful and the people she encounters feel better about themselves. Rather than a fountain, Pearl is a pool of clear water. Instead of drawing attention to her own warmth, creativity, intelligence, and beauty, she reflects these qualities back onto others, whether they can claim these charms or not.

But that is another story. Back to the mustard.

"I'd like to try."

"But why? I'll give you some."

Pearl and I have been housemates before and she is flabbergasted by my desire to make things that normal folk just buy. This is not unrelated to her knowing that I, unlike her, am lousy at crafts. The drapes I make are crooked, the vests I knit unravel, and the broom of dried grasses leaves more of itself on the floor than it sweeps up.

I once brought horseradish back from Illinois where I'd dug up the roots and blended them with vinegar, sniffling and crying through the fiery process. Afterward, I was afraid to open it. Mold refused to grow on it. Two years later, we finally chucked it and the raccoons stayed away from our compost for weeks.

However, the crackers I make are a hit, along with the salad dressing. Lots of the baskets I weave turn out functional, albeit rustic looking, and everyone pretends at least to enjoy my homemade bread. My friend's partner appreciates the sage and witch hazel aftershave that a former sweetie of mine returned after one use.

Still, most of my creations have been disasters. The hat that I knit for my first boyfriend turned out square, a giant tea-cozy. He gallantly wore it anyway, much to my chagrin. One night I caught him stretching it over a basketball.

Why didn't I quit?

It isn't praise or perfection that I'm after. Something in me craves an understanding in my hands of how things are made. It's empowering to know the basics of what I use, to feel that I can provide for myself from my environment.

It's the survival instinct, maybe, or good old self-sufficiency. All my life I've been enamored by stories of people making their own utensils, clothing, shelter, from the basic materials at hand. I'm ever compelled to try.

"Do you have a recipe?" Pearl warily asks as she takes colored pencils to a card she's drawn. The picture is of a small cottage with a large garden, simple but arresting.

"No, but I know the ingredients—mustard seed, vinegar, wine, sugar, salt."

But ingredients don't a recipe make. Mustard seeds float to the top of the vinegar like caddis flies about to hatch.

I give up and we go for a walk in the crisp breeze. We talk about life, poetry, friends as I gather horsetail for making a mat. Pearl smiles and shakes her head. The sky shakes a cheery blue cape that we pass beneath.

The Bath

In the upstairs attic of the main house lies the secret chamber where our most hedonistic activity occurs—the bath. A spacious claw-foot tub surrounded by hand-built wooden cabinets and skylights reclines beside a circular window that frames the trees and fields to the west. A dusk bath allows for tear-jerk sunset views, a midnight one for star-speckled bathing.

Baths, alas, use more water than showers and there's only one tub, so most of us limit them. Our weekly baths are planned, anticipated, coveted. To complicate matters, Jack and Rita, because of their adjacent room, limit our access to the tub god. We decide at tense weekly meetings who can bathe when. The bath quickly becomes a highlight of my winter life at Siesta Lane.

On nights when no one's home, when the cold damp taps insistently on my shoulder, I pin up my long hair and pour almond, lavender, and rose oil into the steaming water, mineral, and sea salts. Beeswax candles give a honeyed light and scent to the room. Sometimes a cloud-petaled moon peeks through the skylight.

In spring magnolia blossoms float on the silken water that caresses my neck as I relax into it, sipping chardonnay in a chilled glass and humming old love songs.

Time drowses while this grateful body sighs and rolls around. Flickering light plays over the oak walls. Steam sidles my hollows, warm currents swirl between breasts and thighs.

When the water grows tepid I dry off with thick towels, rub oils and lotion on tingling skin, wrap in a soft robe, slide into cushy slippers, and float back to my cabin to read romantic poetry by lantern light beneath heavy quilts and the roof's lullaby of rain. And almost forget there's no one joining me in bed.

Love Song of the White Pine

I wait on this green hill cupped
in cloud and ringed by mushroom
where snowmelt gloves
and stirs my roots and wind
trembles my furred boughs
as yellow dust releases
somewhere in the forest bearing
my name. His silk scarf drifts
past spruce and fir to where
my needles flair their skirts and funnel
undulations of pollen that swirl
my shingled cones, and cling
to the rough skin of my scales
for one year before venturing
a messenger to the core,
to spark, to seed, my love
comes to me
in a shower of gold.

Sociality

It's my turn to host the weekly supper club. Every Friday night, a group of writer friends whom I met in grad school gather to swap exaggerated stories and eat decadent food. It helps us get through rough weeks.

I want to hold the event in my cabin this time so I plan a dinner that can be cooked on the woodstove. Cheese and chocolate fondue. Thick chunks of home-made bread and slices of banana, apple, and strawberry for the chocolate. It promises to be a memorable feast.

And memorable it becomes.

Wind swashbuckles through chinks in my cabin and the woodstove goes on strike. We huddle beside a smoldering fire, trying to dip tough bread into coagulated cheese and mushy fruit into lumpy chocolate. The delectable dinner downgrades to "barely edible" and the cozy quarters to cramped.

With no utensils in my cabin, our fingers are soon cheesy and greasy. The festive mood fizzles and in the country, there's no nearby take-out diner to save the day.

"It's better than shoes. People in emergency situations have been forced to eat their shoes," Gabriel offers.

Amy Minato

"Chocolate in any form is still heaven," Sophia suggests, scraping brown mounds off the sides of the pot. Maggie chuckles quietly and chips at some cheese.

We end up back in the main house, where there's a gas stove, sharing stir-fry and Scrabble with Luke and Yoko, affirming the value of community. Raul comes home with new drawings to show and we start a robust fire in the living room.

Maggie stretches on the couch and chats with Yoko, Gabriel naps on the rug, and Sophia reads Raul's fortune in the tarot deck she finds on the shelf. Full, warm, and relaxed, and even though it gets late, nobody wants to go home. The evening has been saved, no, elevated, by my housemates.

Sometimes necessity strips us naked and people pitch in. Being too comfortable and self-sufficient, it seems, may preclude some charming rescues!

Still, it's a risk. The minimalist lifestyle, however good for the soul, can be a strain on your social life.

Cold

When you are cold it's hard to know it will pass, to believe the woodstove will fire, the sun return. It's the terror of inadequate shelter or clothing, and the dependency on our minds and hands to compensate for the absurd vulnerability of our bodies.

At dinner one night we discuss how animals keep warm; fur, hibernation, migration. Mick mentions the woolly bear, who changes its body liquid into alcohol, like an antifreeze, for the season. Raul holds his fork in midair for a moment. "I have an uncle who does that."

We share the cost of wood, gather or hire someone to haul and stack it, but we each have to chop enough for our own cabin. Faced with the need to split wood, I flip through in my mind how Mick taught me to do it—how to hold the axe, position myself, aim, and follow through. I can remember—nothing. Of course, this means it's impossible for me to split wood. Standing here with my fine axe, strong back and hands, good eye, and several hearty logs, I nearly give up ... but I am, darn it, cold.

So in all trepidation I balance a log on the block, swing the axe back and up and down onto it, making a slim wedge in the top two inches. I repeat this many times, slashing and shaving and occasionally knocking the log across the shed. Ceres trots in, pretending to eat from her food dish beside me, sneaking amused glances at me from behind her long ear.

I stumble through the process, bemoaning the forever it takes to chop a single log, getting the axe stuck and hammering the whole thing onto the block until it splinters, somehow oddly attached to the purity of my relationship to this act. Sensing a confidence born of moving from ignorance and failure to, at least, marginal success, on my own resources.

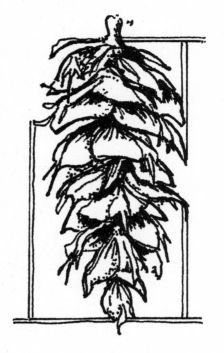

I peer into the woodstove in vulnerability and hope, stacking small to larger pieces of wood, leaving room for air and channels for the fire to reach the top log, admonishing it, in the depth of chill in my bones, to light.

Slowly, almost as an afterthought, the way lights go on in houses across a bay, a few flickers catch. They lick and tantalize and finally flare.

I lean back, squat on the floor, and whisper *gracias*.

What I learn is respect and patience and my own fragility. How much more profound these qualities would be were I even more directly dependent on my natural surroundings for food, heat, shelter.

Sun

As the snow melts I grow soft in my body, gazing through the steaming glass at the flesh and texture of the earth after a week of ice. The landscape awakens, raw, like me, relearning warmth and the green spiriting of the senses. How can I move but from kindness? The grass roughs up yellow and tousled from its scalp of snow. My blood tender as finches quivering on the spruce, unprepared for this sudden thaw, braced so long for cold. I'm wary of the soft opening of soil and what may nudge a pale head up.

In Paris some tourists ask a photographer what they should go look at. "Don't look at anything," he says. "Look at the light."

Today is Winter Solstice, when day and night are equal, and after which days begin to lengthen, the sun returns. To honor this I will move through this day accompanied only by natural light, studying the variations of light and shadow, white nappy bursts in cloud corners, soft shine on wet rock. My body tunes to the slow curl of sun, aware each moment where it pulses, orienting to its path across the sky, shifting my center for its rays.

Amy Minato

Christmas Bird Count

Three of us gather at dawn
two varied thrush beneath the laurel
binoculars in reverent rings around our necks
three California gulls hover and circle
winged and beaked in wool scarves
we scratch numbers with stiff fingers
five rock doves roost beneath the bridge
two winter wren twitter the dogwood
away from task and chit-chat
each of us come to become lost
six mallards maneuver on river glass
five yellow warblers dribble gold against grass

to follow the heron's still gaze upstream
to be caught in the swoop of red-tail, bounce of finch
a ruby-throated hummingbird dives a scrub jay

A towhee bobs a black head as we ponder the pines,
traverse the butte, willing
vessels for birdsong and wingflash,
searching the nimbus, we are Magi
noting the season, and what life
may greet us this day from the sky.

Winter Solstice is my New Year. As Earth tips toward the sun, greeting an old friend, my heart opens to awareness. It's a time to rediscover self, to strengthen gifts. I vow to write more poems, walk daily, give blessing. Crimson lines inscribe this promise across the late afternoon sky.

At sunset I write on small pieces of red paper what needs letting go—and what to wish for—crumple and toss these in my solstice fire. By morning I sense a softening, a focusing of attention, the directing of light within, already affecting change.

January

I look out and see the oak, pine, meadow, and hills. I sit on the deck and want to stay forever gazing at the grass swaying and the mountains turning deeper blue. Listen carefully because that's the whole story. That grass, those hills, that stare, with time filing her fingernails beside me.

Hibernation

The combination of new moon, winter, and menses is redundant and overwhelming. There's no hope of action. There's eating and hibernation and long stares into the fire. There is hot cocoa, sunrise, and sunset.

The call of the body strengthens. When sleepy, I sleep. Hungry, I eat. Stiff, I stretch like a cat on a rug. I curl up and swill with good blood and float through the day, tides in my uterus pulled by the moon. Out go the plans, papers; projects swirl, my boat rocks. My head says finish this, that, call, plan, scheme, scribble, make money, plant seed. Instead I crawl to my cabin to read, doze, and dream, a dumb beast dazed and blush and blooming.

Much of the cold season I remain under cover. Under several covers. A flannel sheet, two serapes, a poncho, a quilt, and a sleeping bag. One of many respects paid to the season is to put out the next day's clothes before going to sleep. This minimizes time spent in the cold part of my cabin, below my sleeping loft. Long baths before bed with crushed sage leaves, almond oil, candles, and poems can last me through cold nights. Long walks circulate blood. Hot cocoa in a bowl warms the hands, as songs do the soul.

Siesta Lane

These small gestures are necessary because cold comes to court us, woo us to numbness, to stagnation, hibernation. If I give in to it, cold can deflate me. I will cry next to a failed fire, sleep early, wake late, eat popcorn for dinner. So when the world is abandoned by the voluptuous eyelashes of the sun and the sky a traffic jam of clouds, I muster a pedestrian courage, the inimitable task of love, and warm the hearth, rub the skin, crack the frozen face into a smile while reaching out for someone nearby to kindle.

Sweat

Raul has decided to build a sweathouse. Everyone is excited to "do a sweat" with him.

For days, he carries wood and blankets, sticks and mud up to a well-hidden spot on the hill behind his cabin. When it's ready, five of us show up in loose tunic-like clothing, feeling both silly and somber. Rocks, heated in an outdoor fire, are brought in to create heat in the small dome-shaped room. We crowd together around the hot rocks. Raul begins drumming and singing. The sound comes from deep in his body and seems to be speaking to the drum and to the smoke. Rising and falling, louder and softer.

Soon we are in a trance of sweat and song. After what seems a long time, Raul picks up a handful of dirt, which is the altar, and sprinkles it on the stones. He says a blessing for each of us and a prayer for the earth. We are invited to speak. Through the dark and heat our murmurs mingle and drift upwards with the smoke.

"I pray for strength with my work."

"For my sister recovering from an addiction."

"Praise to this land."

"I am thankful that my body has healed."

"May my family learn to accept me."
Raul taps us with sweetgrass switches as we leave. Some of us are crying, all of us are sweating. For the rest of the day we are very helpful to each other and tender, speaking in whispers.

Later I reflect on how amazingly comfortable we all felt with Raul's ceremony. If plopped down on Earth from another planet to visit all the major religions with their holy books and icons and elaborate temples, I would find myself drawn to Raul's altar of soil. What, after all, could be more clearly, unabashedly, worthy of praise?

Compassion

My friend at the magazine is swamped in submissions so I end my leave of absence and drive out to the truly intentional community where it's housed, to help.

Skipping Stones operates out of Cottage Grove, a community about thirty miles south of Eugene, up a long rutted dirt road, that's way beyond Siesta Lane on the sustainability scale. Aprovecho, which in Spanish means "to make the best use of," is a collection of folk fiercely dedicated to conserving resources. On my quest for simple living, I had briefly considered living here, but as someone who barely survived the small sacrifices of Lent, like giving up candy for forty days, I knew I lacked the stamina.

Most of the scattered buildings have no electricity or water. One's made of mud, another of straw. One's thirty feet up in a tree. Aprovecho folks are dreamers and inventors, young people with cold cheeks and old clothes who do things like give classes on cooking with cow patties. They travel across the world to build, alongside villagers, rocket stoves that use very little wood. Rocket stoves help villagers both cook and heat their homes more efficiently, and conserve their nearby forests. At home

131

in the United States, these workers live much like the indigenous people they help.

Hot water at Apro involves stoking a fire first, or waiting for a sunny day to heat the solar shower. Food, almost exclusively, comes from the garden and bread is made by hand. If you don't like steamed kale and fava beans, you'll lose weight here. Many do.

Every year a troupe of new interns comes to learn permaculture (the art of enriching soil through farming and ecological building) and to experience living in an "eco-community." Some can't take it. Others become enchanted and never leave.

When I arrive today for work, two stocky men with knit caps are designing a solar-powered refrigerator in the shed. One's laughing loudly and the other grins. Friends for a long time, and bearing a not so vague resemblance to Tweedle Dee and Tweedle Dum, we call them the mad inventors. They've already designed a solar-powered desalinator, which a coastal Mexican village now uses to make seawater useable for crops.

Hens run back and forth along one row of the immense garden in their tubular wire cage, dubbed a "chicken tractor." The tractor gets moved along the rows so the chickens can fertilize the veggies and eat the slugs across the entire garden. Residents in turn eat the chicken eggs.

Gravity carries water through pipes from the spring on the hill to the garden. A goat near the main house keeps the yard trim, and the many compost toilets provide organic matter for trees and shrubs.

Garlic and herbs hang from the ceiling in the main house; sliced pears dry on racks that swing above the woodstove.

I work on *Skipping Stones* magazine with Asit, whose tiny cabin nests beside the garden. To get there from the main house I tightrope over a rickety plank across a creek and pass below a thirty-foot-high treehouse through a field of blackberry brambles. This gingerbread house is filled with files, magazines, submissions, and subscription materials. The floor-to-ceiling south-facing windows heat the rock floor on a sunny day.

But today we can see our breath as we work, and even though I'm charmed and inspired by Aprovecho, sometimes the mud and chill have me longing for an office with bright lights, thick carpets, and central heat.

True to Asit's East Indian custom, we drink cup after cup of rich, creamy, spicy tea. My spirit stirs with the writings from children around the world:

My Father's Illness

Yesterday my father
went to carry hay on his back.
He trembled
as he bound the hay into bundles
and carried the bundles on his shoulders.
He took a break on the hill at Sung-Choon's house
and as he was getting up
he bumped his head on a rock behind him
and fell
rolling into a ditch
with the bundles still fastened on his back.

A stranger came to his rescue,
supporting him
as he sat there gasping.
His breath stopping
and starting again
over and over.

Siesta Lane

I went there and wept
looking at the height of the hill.
—Kim Kyu-pil, grade three
Andong Daegok School, Korea

Hearing daily from children living in much harsher conditions makes my material sacrifices easy. If I wasn't awash with their stories, I could pity my poverty amid the riches of America. As it is, I'm blessed with shelter, clean water, abundant food, and libraries.

Young, hopeful voices in many languages strum inside me as I edit submissions until we lose our working light, the sun.

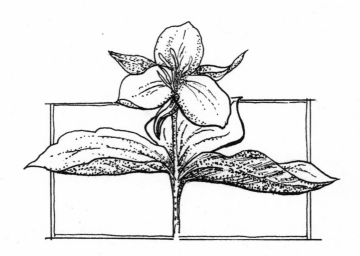

Tenacity

A teen girl with two nose rings and a tattoo of a skull on her arm wrestles with the front tire of a bike on a stand. In this Santa's workshop, she's rebuilding a bike for herself.

The Center for Appropriate Transport (C.A.T.) in Eugene, where I have just taken a job as a bike courier once a week, hosts bike labs for youth where they receive school credit for learning shop skills. The kids can keep or resell the bikes they build. On weekends, they help out with bicycle valet parking at large city events and split the profits.

On the surface, the director couldn't be more unlike Santa Claus. Tall, lean, and striking, Pete has hazel eyes that flash with intensity. Pushing a welder's cap up from his sooty face and waving a soldering iron, he rails about the increase in traffic in Eugene, the grease on his work clothes not unlike blood on a warrior's armor.

He's on a crusade against cars and has convinced the city to take heed. Partly because of him, each year the city has more buses and bike paths, and less pollution. Pete builds custom-made bikes for people with disabilities and cargo needs: bike taxis, recumbent tandems, adult trikes with baskets, hand-powered bikes. His wonderful inventions breeze by, reminding us there's a wizard in our midst.

C.A.T. runs a co-op where members use the shop to fix their bikes. Pedaler's Express, a bicycle courier service, is also under C.A.T.'s umbrella. These couriers are not daredevils like the city bike couriers. With big cargo cabins in front, they're lucky to keep up with other cyclists on the road.

On my shift, I ride for six hours across the city, rain or shine, delivering photos, legal documents, campus mail. On dry days I can't believe they pay me to do this. Breezing beside the river beneath trees and birdsong, knees and blood circling, dreamy. College students stop me to find out where they can apply for such a sweet job.

But during icy winter rains it's another story. When I get a flat nowhere near shelter, already late for an important delivery, and my frozen fingers can't get the tire off, no one wants my job. I don't want my job. When I finally stumble in, the clients offer me hot cocoa as I sniffle and drip on their carpet.

The three upbeat muscular women who do this five days a week, all year, are Goddesses of Social Change. I know myself to be a diffuse activist, helping here and there, careful not to bear the full weight, nor to be the final authority. Ashamed at my half-hearted dedication, I tell myself that my primary work as a writer is to witness and retell, to keep the rest on the periphery. But the real reason could be fear.

Maybe fear that Pete's fierce commitment to a cause would swallow and devastate me with each setback. Fear that caring about and identifying with something so intensely might strip me of the freedom to change.

Amy Minato

The teen has gotten the front tire off and now sits outside smoking a cigarette, looking surly. She dabs the sweat off her brow with a crumpled bandana. I imagine she is deciding whether or not to bolt. She could renew her life of independence and danger on the streets or keep working on her bike.

She smashes her cigarette on the pavement, picks up her tools, and goes in to grease an axle.

Wanderlust

It's been over a year and I'm missing my jet-set companion today, slipping into desire. Tall, sleek, and gracious, Alex enters a room like a prince from an exotic land, magnetizing, but mysterious as a cat. From humble college beginnings where we turned my dorm closet into a tea room, life with Alex became a cultural feast: big cities, foreign places, pomp and pizzazz. But throughout his catapult to architectural success, Alex retained the calm dignity I first loved when we met in college.

Ambitious and talented, Alex introduced me to Hong Kong movers and shakers, an eclectic group of expatriates. Although charmed, I sensed a restless loneliness within them, a homesickness that could have been a projection of myself in their situation.

"Ghost people," the Chinese call white folk. Maybe in addition to our color, it is because when they see us in their country we are disembodied, removed from the flesh of our homeland.

Alex (or Pui, his given Asian name), a one-woman kind of guy, patiently explained to me, his flighty friend, the merits of devotion. His career promised a rich life of travel and adventure, his company, a refuge.

Amy Minato

Scared and confused, I shrank from the challenges and isolation of a life spent in hotels and restaurants. This gal needs nature and community.

Although devoted to each other, we couldn't reconcile our urban/pastoral split. After twelve years of visits and long distance phone calls, we set ourselves free.

Still, when mist turns Rainbow Valley into a Chinese ink drawing, and jasmine tea brews in a flowered pot, I remember living with Pui's family in Hong Kong—going for late-night jok (rice porridge), slapping mah jong tiles, passing the orchestra of tai chi enthusiasts in the park at dawn. I had to nap frequently from the stimulation of that pulsing city.

So here I remain looking out on a meadow instead of an international kaleidoscope. If I had married Alex, where would I be right now? In a Japanese bath house? An Italian plaza?

If I hadn't already been that route, and known the dizziness and alienation along with the thrill and intrigue, I could feel trapped and narrowed at Siesta Lane—the same old muddy rug, the same old loquacious rain.

Instead I miss the man and not the lifestyle, knowing that most of my habitat needs are beautifully met right here where I'm planted.

February

Tired and in awe of the terrific work, I peep in at the room filled with straw. My eyes ache, tear up with the thought of it. I back away and enter only moments a day, turn the spindle, and get carried in the motion and twirl of threads pulled into sentences and long silences, hoping to spin a few tangles of my life into gold.

Resonance

Something's up with Luke and Yoko.

She looks pale and he hasn't shaved in a week. When we greet them, their replies are terse, eyes averted.

Because of the shared bathroom and kitchen, one has solitude but little privacy in our curious community. It's not our business, but we care about Luke and Yoko, so we mull over these phenomena.

"They're having a fight."

"She's seeing someone else."

"He's moving away."

"She's pregnant."

"He's wanted by the FBI."

"What?"

"You never know. ..."

We never do know what happened, but Yoko stops coming home with him. They were hit by one of those unexpected waves that send relationships into the Bermuda Triangle of Love, maybe. For several days, Luke, red-eyed and weary, sneaks dinner to his cabin, alone.

We all feel strangely lonesome and miss Yoko's earnest cheer and exotic clothing.

Siesta Lane

When you live with folks, their moods rub off on you like dew. This can be exhausting, exhilarating, confusing. Some days you're not sure whose mood you have, why you feel jazzed suddenly, or spaced out. But if you look around, you know someone's new job has created a stir, another's upcoming final has us all on edge. Each of us vibrates with the group's emotional tunes.

We know it's getting tight when we begin having each other's dreams. One night, I am painting a series of mole tunnels. In the morning I give Raul back his dream and ask around for my own. Sara offers one about surfing that Luke claims, Natalie snatches up Rita's horse fantasy. No one takes Paul's nightmare about large dogs.

Sometimes over breakfast we barter our dreams, embellishing for effect and accepting pledges of waving bagels from the highest bidder.

Dance

This winter I have discovered the irrefutable genesis of all human dance. We must have invented it to keep from freezing our butts off.

On nights when the damp northwest cold seeps like a rumor through my bones, when I'm trying to cook a quick, late dinner in the main house but haven't had the patience to make a fire for the warmth; or when there's no kindling and my attempts without it have resulted in sad piles of smoldering newspapers, I move. To keep warm. And to keep from getting cranky.

For variety I move in rhythm, in different ways, to warm every cold crevice of my body. I put on Motown or Lyle Lovett and tap to the slicing of potatoes, sway to the onion sauté, and go wild during the veggie simmer.

Pretty soon I've taken over the living room, flailing and gyrating, toasty and laughing and the few folks who come in for a cup of tea discreetly change their minds.

I eat my slightly burned dinner with a flushed face and heart and think, with profound awe, of our tribal ancestors in cold climates living in thin shelters and sleeping darn close to the frigid ground. Did they envy their furred companions? Did they move in rhythm around a campfire to keep from freezing? The genesis of dance, then, may have been a triumph of the human spirit over the humbling restrictions of the flesh.

Silence

There is a quiet stillness in the country that invites reflection, intuition, and spiritual growth. What happens to an individual, and a society, denied calm and contemplation, tree-climbing and rock-skipping?

Now I notice the spider in the hall pocketed in its corner, legs tensed, sensing any pull on its web, the way color refracts off its tear-shaped body, the delicate threads of its home. Since pondering the precariousness of life suspended in a translucent net—so invisible yet so present— I have grown to care about the spider in particular, and in general, and she in myself who weaves and waits and senses which beings entering this web of memory and imagination are nourishing, and which are not.

The silence of Rainbow Valley can be both soothing and terrifying. Without the clatter of cars, lawnmowers, and televisions, I am left with my heartbeat. Serenity peeks a timid head into my life, but is pushed aside by specters. Alone with my mind, it's hard to quell the voices that play back precise descriptions of my flaws and failings. It becomes strangely appealing to turn up the radio, buy a TV, or carry on inane conversations.

Siesta Lane

Against all training, I turn a rusty ear within, to listen below the din.

Soothing and hypnotic this silence, a soft bed. I begin to crave it, sink deep in and pull the covers up as layers of trivialities melt away.

I go hours, days, without speaking, digging into my burrow of silence. The beings out here seem without guile, malice or pretension. Grasses grow up toward the sky, breezes blow and quit. A robin fusses with her nest as I shake out a rug. A squirrel schemes a way into the bird feeder. Raccoons scurry defensively across the yard. Recognition of the parallels between human and non-human animals nurtures a fondness, a sense of family, blood relatives.

Rabbits and squirrels take notice of me. They recognize this laconism that allows one to concentrate, remain hidden, and conserve energy. For me it preserves my dignity. There's less chance for misunderstanding, or for trying to say something large in words that are too small.

But silence's greatest gift is the honing of other senses, even the hidden ones. Every red hue in Luke's  hair shines at me this morning, each crackle of the fire a different tone. The rough weave of the placemat tickles my fingers, coffee scent permeates and lulls.

The power of prayer may be its requisite silence.

At day's end I sit quietly in the corner while my housemates worry, whispering about me as if I were not only mute, but deaf.

"She's been like that all day."

"I wonder if she feels okay? There's a nasty flu going around."

"I think it's a sign of depression."

"Maybe she needs a date."

Siestans don't quite accept my explanation that this is an experiment, "a writer thing," figuring it's a cover-up for a bad mood. Meanwhile silence unwraps gifts inside me; and the chickadees out the window carry my heart through the branches in graceful, delicate sweeps.

Quiet is best experienced when accompanied by "sloth." (English has few words for inactivity and they all have negative connotations. As a culture, we loathe doing nothing—it's not productive. For relaxation we exercise, shop, watch television.) I remember watching the old men in Greece, sitting in the outdoor cafés at dusk, down by the Gulf of Corinth, empty ouzo glasses, a few olives, and the remains of a squid snack on their table, dark eyes on the darkening water. They knew it depended on them being there, doing nothing in the dusk, for the Delphi ferry to arrive, for the stars to come out. If it wasn't for their silent nightly vigil, their tiny town might slide unnoticed into the sea.

Likewise, would Siesta Lane, as reflected and rendered here, disappear without my rapt attention during this contemplative pause in the whirling center of my life?

Fecundity

New lambs slip onto wet February fields slicker than the mushrooms that sneak up in the fall. Suddenly, they dot the landscape with their furred blur. Wet and tottering, heads hovering over knobby legs, the lambs haunt their mothers, slipping mouths to udders at every lucky chance.

I knit and sing, bake and clean. My hands linger too long over the cat's nape. My womb stirs and clamors, wanting a child. This desire is deep and familiar, as strong as that for a lover, but more urgent.

What is this maternal urge—a trick of biology? Or the soul's need? The earth suffers from a surfeit of people yet I want to produce more. Isn't such desire an unfortunate relic of human evolution, not unlike an appendix or a wisdom tooth? Too many of us certainly can't be good for our species or for any other.

Tears rattle me anyway and I rock in the oak, wanting, wanting, children. Babies visit my dreams knocking to enter this world. They seek my body even as my optimal childbearing years disappear one by one like bunnies in a burrow.

Amy Minato

I'm stumped by this one. I made living near nature and choosing meaningful work happen. But finding a mate and fellow parent takes time and luck, neither of which appear in abundance these days.

I review my list of suitors over the last few years—the one who never kissed me; the one who wanted to kiss me all over, too soon; the pathological liar; the zealot. Nice guys to hike with now and then, but live together? Raise a family?

The robin collects moss for her nest. The fox lines her burrow. I flip dejectedly through the personal ads.

Hot to trot stallion lusting after magical mare.

Gay man desires serious commitment.

All brawn and no brain seeks same.

Maybe I'm too picky. Spoiled by long association with a good man. Alex, after all, is a hard act to follow.

With no compatible partner in sight and without the gumption to bear children alone, I ache and consider giving up this dream, while the hills offer plump bellies to the sun with each turn of the fertile earth.

Fear

I walk, often, at night down Siesta Lane. Sometimes with a knotted willow stick, armored with years of precautions, shedding anxiety, seeking connection with the nocturnal. A form of fear therapy. Only it is not the dark land that scares me, but passing cars, flicking their brights at the roadside phantom, and the sporadic rash of dog bark.

The vagaries of our natural environment used to be the focus of people's fears—drought, cold, wild animals—now we fear ourselves. Every morning we open the newspaper to horrific stories of domestic abuse and random violence. A teenage boy brands his sister with a hot iron, a couple "collects" young women and kills them. Often such news comes from far away, as if the journalists scoured the globe for the most shocking stories they could find, rather than writing about some local folks' daily struggles. Thus we breed suspicion rather than compassion. We build more jobs, hire more police, covet our stuff. Our society's penchant for the dramatic may have us missing out on the quirky good humor of a fellow passenger on a bus or the shifting hues in a lingering sunset.

152

Amy Minato

Tonight small creatures scurry in the brush. Bats—these first "witches"—the only mammals that truly fly, wisp among the winded trees. Using sonar to navigate, they bounce high-pitched sounds off nearby objects that "echo" back, giving the bats exact positions of their prey. In China, bats represent health, happiness, longevity, and peace of mind.

Eyes relax and expand at dusk. Pores open. Mouths and faces soften. Scents sharpen. Pine sap. Sweat. Musk. Sound carries in the still dark. Feet shuffle the path. Smell and touch and hearing are the channels of the sensual night.

I stop by the horses. Comforted by these shadowy mounds bending their heads to the earth, their generous animal smell and rhythmic munch. The three chestnut mares look up and wait, their manes beige halos, ears up, nostrils wide and soft, sensing me.

Horses have a guarded aura, vacillating between pride and terror. They have been domesticated but not slaughtered, caressed but not pampered. Through centuries of close association with humans, it seems, they have maintained dignity, proved their power, earned respect, kept their strength.

This late winter night I turn off Siesta Lane and startle a young stallion in the middle of the road. He has jumped the wire fence slumped around five acres of grazing land. The land so ample and fence so thin as to give an illusion of liberty. But maybe long subjugation has created in the horse a psychological cage, a dependency on limits imposed by humans. What family expectations and social mores circumscribe my own life I wonder? And how solid are these boundaries? Could I risk freedom over security and stand to jump my own fences?

The horse stands regal, but paralyzed by indecision, flanks quivering with fear and freedom, eyes pleading for relief from this strange panic. I could chase him away from his owner's land, to a wild existence with other feral horses and possible winter starvation, or lead him back to pasture.

My own inner pendulum used to swing wildly between freedom and security. One day I'd look at travel ads, the next at real estate. These days I seek a cautious freedom, an open security.

I take the stallion's halter, walk him back through the gate, and return home slowly to my own familiar restrictions.

Darkness

It has taken months of living here to shed my anxiety about being a woman out alone at night. But finally, now, like a nocturnal animal, like a deer or a bobcat, I leave on my night walk and enter darkness as one does a familiar room. Feeling safe, hidden, my eyes rest in shadows and my pace slows.  Trees wave their gentle hands, plants release their green scents, and the low call of a great horned owl echoes through the chambers of Rainbow Valley.

This darkness feels like a presence, palpable, filling the spaces between trees and buildings, rubbing against me like a cat. Night light is receptive, resting on its haunches, the realm of dream and angel, ghost and vision. I go into darkness—this fertile place where people deepen, grow, make love, give birth—for refuge.

In Defense of Defense

Of course the robin hides her egg
for its blue is nothing else the color of.
And the western pond turtle must
pucker its soft neck
beneath the clammy rim of its shell
and the reclusive beaver slap and dunk
the poppy purse its petals
and scotch broom pollen
steal onto the leg hairs
of the digger bee.

It's obvious why the butterfly fins
of the rainbow trout waft
in the speckled shadow of its stream
and the coyote trots to the sheep farm
always in pith of night. We know
how perfect is the dead leaf color
of the wolf spider and the fake sleep
of the garter snake surprised on the path.

Because some memory shivers
in the chamber of our cells.
Because we too harbor
the slug of our heart
from what may apprehend it.

A car rounds a curve and its headlights stab at the land. I slip behind a tree, into its safe shadow, and look up to where a great horned owl, coddled in an aura of bronze leaves, blinks back at me.

March

I've stopped being pretty about it. I leave the window open for moths and examine dung. I take feathers from dead flickers and pheasants. Spider webs cascade the wall. I study the skin of roadkill deer, watch compost decay, ants fornicate, my cat, Quixote, chaw on a baby mole.

Spring

Silverspot

Somewhere a male silverspot butterfly
perches on a lotus of light, waits
for a female to flicker by
so he can dance with her
in a spiral of wing and air
nectar dripping from her proboscis
pollen off her legs.

When male butterflies come
he frightens them away
ferocious in his habitat of sun
place moving eastward
through the day. As our lives move

always in the direction of desire
the shifting territory of what we hope
to own. Does the butterfly
behave thus because he knows
he will die at summer's end
or because he doesn't know?

Amy Minato

Caddis fly larvae collect pebbles and glue them together into little rocky cocoons inside which they transform into their adult form. In March, after one year's incubation, the caddis flies muscle up from river bottom mud, float past the swollen mouths of trout toward the translucent rim of water before sky, and explode into wing on the riffles. And, in her two weeks of celestial life, each female dances aerially with a mate, dives back through current and sea- 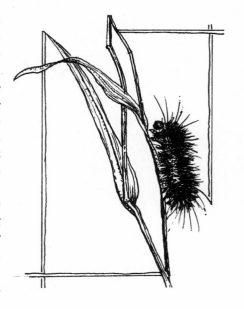 weed to bury her eggs back in the mud, herself in the river.

The painted lady butterflies migrate through Lane County this month, coming from Mexico en route to British Columbia, heavy with nectar and new wings. Mick was driving home on West 11th, a strip of stores and signs, when a cloud of them tumbled over his car hood and across the window, fluttering and turning, fly- ing on instinct and faith, mistaking the bold red on the billboard cowboy's bandana for flowers.

At Siesta Lane spring creeps in the back door and hangs up her muddy parka before we know she's here. Suddenly the field is freckled with daisies; lambs are bleating on the neighbor's farm; and the wind is a teas- ing pat instead of a wet slap. The men wear bright shirts and the women cut their hair.

Siesta Lane

Restlessness shakes its rattle through the group. Raul disappears for a few days and Luke plays loud salsa music; Sara joins a Frisbee league; Jack and Rita cover the living room floor with maps of Europe. Mick meanders the valley seeking rare plants and Natalie and Paul order seeds. I, of all things, meet a man.

And after so many cold, introspective months love is sweet nectar. I get drunk on the woozy-woo.

"You're in the love shack," my friend Pearl observes.

And it's all about that place, this time, not about the guy. The pink center of the rose into which I stick my head until I nearly suffocate.

Riley is a sweet guy with hair in a cloud of dark curls and an uncanny resemblance to my very first boyfriend of the same name. A rare book curator, he's crazy for fly fishing and experimental jazz. I can generate no enthusiasm for either of these. The first makes me squeamish. I manage to avoid actually catching a fish when we go. The second, when listened to incessantly, gives me a headache.

This fellow is rabbit energy—fast and wary. He practices meditation and tai chi and hangs around me to slow himself down. But there's a string in him drawn so tight, that, when plucked, sends him to another planet. I witness this a few times to my growing bewilderment.

The relationship is doomed but irresistible. It plays itself out like the caddis fly—brief and fatal. I'm left dizzy and chiding myself, humbled by vague longings, unsure of my instincts but drawn back to my craft, my community, and the land for sustenance, waiting out the season.

Gatherers

Most of us shop faithfully at secondhand stores, getting good at finding nice stuff in decent shape. We like knowing that no new resources were consumed just for us, that no sweatshop gets our support, that some other mysterious person once appreciated whatever we buy, and that we spend one tenth of what we'd pay for new stuff. Although I used to earn twice the income, I now spend five times less. So I have fewer debts and more freedom, if less class.

It's a treasure hunt! Aladdin lamps, Jesus candles, ships in a bottle, all the basics. There are usually kids trying on funny hats, reading torn but legible books, or playing with toys that someone loved into their current conditions. It's a friendly refugee camp. All of us jetsam from the fast pace and high cost of modern life.

An Ecuadorian environmental lawyer visiting Aprovecho comes along with me to a thrift store. He stares for a long time at the price of cotton shirts, wondering if his English is failing him. They are a fraction of their cost in Ecuador, and this is the expensive United States! Shirts for two dollars each! Louis begins pulling out every usable man's shirt in the store, thinking,

I'm sure, of seeing all his friends back home in "new" shirts.

He begins in the blue section. Every once in awhile I look over as he starts in on a new rack. He methodically lifts shirts off their hangers and fills cart after cart like an assembly worker pulling bottles off a shelf.

As we go through the checkout line with our seven carts full of men's shirts of all colors and sizes, speaking Spanish to each other, the cashier cheerfully asks him where he is from.

"Ecuador!" Louis answers, smiling, stuffing shirts into bags.

"Oh," she replies empathetically, "I guess you probably couldn't bring many of your own clothes with you?"

Yard sales, although offering few items, simmer with stories—the residue of someone's life, that person sitting in her garage amid the hand-knit Christmas stockings, two for a dollar.

You look at something and the woman standing guard over her goods starts talking.

"My husband bought that canoe for us before he died. We were going to take it to Canada."

"You can have all those canning jars for five bucks. Can't do preserves anymore with my bad wrist."

I look at this woman wizened by life and take her home with me, in empty jars that fill with spiced applesauce, stewed tomatoes, awe. And there is gathering in the more traditional sense. In early fall and spring my friend Willa, who looks like a wood sprite, and I, go gleaning. We talk and meander through lush woods, eyes open for finds. We glean wildflowers for salads, mint for teas, mushrooms for quiche, vines to make baskets, and the indomitable stinging nettle for soup.

Nettles have been harvested for food since ancient Greece, and Germans today use them as a remedy for arthritis. Their limp, pale, tear-shaped leaves are some of the first to appear in spring. To gather nettle one needs long pants, gloves, and faith. Long pants and gloves for the poisonous hairs that sting your flesh, faith that these stingers will be rendered harmless when boiled—which they are, honest.

We cut the young, new nettles, clean them, and drop them like lobsters into boiling broth. Tasting not unlike spinach, and as my new friend Joseph points out, not unlike mowed grass clippings, stinging nettle is laden with calcium and iron. After eating the cooked nettles drenched in olive oil and soy sauce, our faces flush with their potency.

In late summer we collect rose hips, the shiny red beads that roses turn into, for their vitamin C. Rose hip tea is a favorite, with its tart, berrylike flavor, though Paul chews the hips like gum. Rose hips strung together were the first rosaries, as roses are associated with Virgin Mary. I try making one for my mother, but it quickly falls apart in my hands.

Even more than thrift-store hopping, wild food gathering gives me the sense that I am in some control of my survival, puts me in touch with my ancestors, and prompts me to study the local natural environment. Sometimes, in the case of certain fruits and mushrooms, it's even delectable.

Acuity

"Sounds like a robin."
  "But more melodious."
  "Well, then it's an evening grosbeak. Anyway they should be arriving around now."

Precision

How can the violet green swallow
swoop and catch in beak the curl
of floating feather, and the beaver
chew on the ash just so
it tips and slants into the stream,
and tree frogs chirp and halt
in unison, and what makes
the shooting star's flower the same length
as the rufous hummingbird's thin beak,
and how do the eight eyes on the spider discern
on which chip of bark to press
its spinneret's wet tip, and why do
the blue-capped stamens of the pale flax huddle
so symmetrically around its pistil's cotton top,

Amy Minato

and the sun leave a gold egg on the altar
every solstice, and how can each tiny
winter wren repeat so elaborate
a song, and lovers fall into each other
so deft, and so dumb, and especially
how can the rose campion be that exact
magenta every time?

A year ago I wouldn't even have recognized a robin's song, let alone a variation of it. I didn't know anything about birds, which ones appeared when and where. Flora was another area of ignorance. One yellow flower looked like another, pines were pines. Into a poem I'd toss flower names like lilac or narcissus that had literary allusions or nice sounds but that were completely out of context. Since turning my attention to the land, I have learned to distinguish a lanceolate from an elliptical leaf, a damselfly from a dragonfly, the call of a barn owl from that of a great horned.

My senses are whittled into finer form, honed and sharpened on nature's whetstone. With my first pair of glasses I remember being startled by the crispness of the world lost to me for eleven years. Now, whenever I learn a new bird or plant in my daily realm it seems incredible that I didn't notice it before. The world rushes into focus, turning those comfortable blurred edges into leaf gall, eye ring, seed head, spider web.

Making such distinctions about my environment gives me immense satisfaction. So if I have trouble deciphering computer lingo or affording life insurance ... at least I know my neighbors.

Wind

The breeze has gotten beneath my skin. Veins jiggle, sinews swirl. In Spanish, you don't use a pronoun when speaking of your body parts. You refer to "los piernas"—*the* legs, not *my* legs. Indeed when the wind coils I can claim neither limb nor hair as my own—harkening, as they do, to this new master. The heart beats faster and the body responds to a visceral language with feelings from ecstasy to terror.

Wind is seldom welcomed in city life. It scatters papers, messes hair, drops debris across the cultivated lawn. But with no appearances to keep up, I'm mesmerized by wind, coming as it does from the ocean's gyres, brewed air, the moon's pull.

Sirocco—a hot, dusty, humid southeast wind. Zephyr—a light, warming breeze. Santa Ana—a strong, hot, dry wind. Mistral—a powerful, cold, dry northeasterly gale. How many words for wind? What do we call the breeze that lifts the tiny hairs on our cheeks? The one that brings trees crashing to earth, or makes hieroglyphics on the lake's skin? How different is a blossom-scented breeze from a leaf-ravaging gust? We need words for these variations, these gestures of air. We need a way to talk back.

Wind brings whoosh and ripple, snap and sigh. It highlights the dynamic earth, shaking it before us like someone with a lucid article. "Here, read this," wind says, "these branches scribbling sky, the calligraphy of grasses." Or, rattling its morocco, "Listen, the world is a marimba band. Get off your chair. Go outside. Dance. Dance with the land."

And so I do.

Reciprocity

There are many relationships with land. Nature as provider, mentor, lover, friend. But with a tweak nature may seem like withholder, beguiler, betrayer, enemy.

A tummy warms with soup from veggies grown in our garden, clean air laps lungs, water cascades through veins. Nature as sustenance ... or destroyer. When crops fail, floods rise, or flesh freezes. In our cocooned society we sometimes forget our dependency. How humbled by disease and natural disasters we become. How cocky we can seem. But in our bones, we know it. We buy insurance, store up food, and watch the sky.

A heron, still as slate in a river, teaches patience. A jay scolding my cat, courage. I remember, during a particularly tough period at the magazine, walking by the Willamette River in Eugene. Every bone and muscle sent memos to my brain, requesting vacation time. But we had a deadline in two days and I had to postpone time off until then.

In the cold grey mist I had discerned two teeth and a large stick moving across the water. A beaver was pushing a log against the current. It took no notice of me, but I strained to keep it in sight. So sleek, so slow. This diligent critter that has carved landscapes—creating wetlands, controlling floods. All that week I dreamt of the beaver. As I leant over a desk crazy with papers, rising early and working late, the image of that dark wet animal steadily moving its heavy load forward etched into my memory. Almost subconsciously, laced with gratitude, I had learned endurance from the beaver. Its image inscribed into my consciousness, like a template of a way to be in the world.

Rain pools in a pocket of soil, a deer with luminous eyes lifts a slender neck from a stream. Quivering, she poises to flee. Bare sienna madrone limbs curve through the woods. I tingle.

Stopped on a walk by an old cherry tree towering in bloom. Delicate white blossoms lace its thick black bark. We're flung from our bodies into the ether of great beauty, and shaken. The Earth as lover: sensual, erotic, stimulating.

Yet increasingly in this midseason of my life, the Earth coaxes me to a more companionable love. Toward a love between friends. I'm called to a reciprocal relationship rather than one of awe and gratitude. This symbiotic acquaintance compels devotion, caretaking, active concern. It implies that land requires, as one friend to another, that I be honorable, helpful, humble. That I give of myself, and consider the future ripples of even my slightest act.

April Dreamt that I recognized a plant growing in my room in a book of rare plants. It was a treasure, and I'd had it all along in my room, not suspecting its worth. It had pale, round, flat, heart-shaped leaves with tiny crimson berries. I understood it was my task to protect it.

Simplicity

Because we can fit exactly one percent of our personal items in our 150-square-foot cabins, we are finally clearing out the barn and sending a caravan to charity. The main house is roughly 1,200 square feet, but we can't put many personal items in there.

Many of our things have been in boxes in the barn all winter, attesting to the fact that we don't really need them. The motto here is: if you want to clear out your possessions, move into a place the size of a walk-in freezer. This goes against current trends.

The average U.S. family size shrinks while house sizes grow. People then feel compelled to buy more stuff to fill up the space, stuff that requires either time or money to care for and store.

Sara, Rita, and I agree to have a clothes swap. Traditionally, at such an event, women ravage like wild bulls through each other's things and what you come up with, if it's not too badly torn from the struggle, you get to keep. However, since the three of us wear sizes about as similar as the three bears, the best we could do was admire each other's taste and tell about our lives.

"Gramma knit this for me."

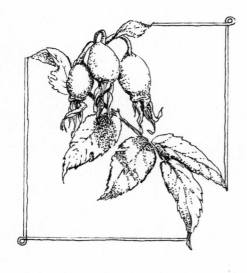

"My first boyfriend gave me this sweater. He had blue eyes and a brain the size of a spitball."

"I got this skirt from a market in China. The old woman selling it was smaller than a child, wizened as bark."

As our stories unravel, I begin to comprehend the attraction of quilting. The way women save scraps of clothing from their lives, remembering the dance they wore that dress in, which child wore that sleeper, the shy cousin who gave them that shirt. How they take these scraps and create art that provides warmth for their families. How, with this skill, nothing is lost, only passed on.

In this same spirit we three women share our stories and pass along to our community in the form of clothing those bits of our lives that have perched awhile on our bodies, and sung to us.

Mud

"Let's do a mud bath!" Natalie suggests. "I hear it's good for your skin," she adds, in a more serious tone. I gaze across the table full of seed packets at Natalie's eight-year-old skin, gleaming with every possible variation of the word *fresh*, and agree.

"Yes, it's absolutely vital for every woman, darling," I respond. "Anyway, it sounds like a gas."

So we wait for a warm day when everyone else has gone to town, pick out a secluded place, and start digging. After we've dug a bathtub-sized hole, we carry buckets full of warm water from the kitchen and pour it in. When the consistency becomes that of brownie dough, sticky and pliant, it's ready.

We find a spot sheltered from the road and strip in the sun. "I hope no one's bird watching," Natalie giggles.

"If they are," I say, "we'll stump 'em. Let's see. One bare-breasted ruby crown and a long-legged mud flapper. Lifers!"

We slide in and cover every spot of flesh with ooze, even squish it into our hair, singing, wallowing, painting ourselves chocolate. It feels sexy; it feels like what e.e. cummings described as "mud-luscious."

Siesta Lane

A car going by slows down, so we slink farther in until it seems as if only our white eyeballs are showing, along with some clumped-up hair. This, of course, makes us laugh so hard we slip getting out. Once in the hot air, the mud dries quickly and cracks with every move. "Ouch," I say, "Youch." Natalie agrees, because grinning is starting to make our faces hurt. We hurry home like two sleek otters or two large earthworms or two muddy rakes, to rinse off.

Afterwards we exaggerate about how glowing the other's skin seems, and how soft! Fussing and clucking, all the while remembering why we really did it, for the slither and spank, the daring, the togetherness, the warm muskiness, of our earthen baptism.

Loyalties

My cat presents me with a moral dilemma. The more dedicated I become to preserving native plant and animal life, the harder it becomes to ignore the disastrous effect pets can have on them. Cats especially, decimate bird populations. Whether or not you feed or bell them, cats tend to be undeterred predators. Keeping them inside is the only solution, an unacceptable one in our case.

After a single day locked in my one-room cabin, Quixote is ballistic. I return from a walk to find yarn strung in knots around the room in a giant blue web that takes an hour to untie. Clearly, the call of the wild beckons. I bring tuna and catnip, try to pet Quixote but he skulks off, scowling at me with low brows and piercing green eyes. My cat is not a meower, but his silent glance is feline eloquence enough. Reluctantly, I leave the window open again.

Each morning this week, I wake to a scrub jay's scold. Investigation reveals Quixote hunched patiently below a low limb on which the raucous jay perches. Clearly, he is the object of her wrath. My cat tilts a bemused head, passively accepting the scratchy shower of "yaaks." Of course animals eat each other, that's natural, but domestic animals have an unfair advantage. They don't need to kill for food, and introduced species can harm native ones.

Western scrub jays are important to the ecosystem. They groom mule deer by eating parasites off their fur while the deer stand patiently still. Rainbow Valley is oak country. Unlike jays in pinyon pine country, where seed-getting requires finer bills, Western scrub jays here have stout, slightly hooked bills to hammer open the acorns and rip off the shell. They will hide acorns, and if another jay is watching, dig them up and bury them somewhere else. Acorns that they don't dig up may grow into oaks. Mutualism.

Scrub jays are also smart and bold. This one, obviously, has a thing about my cat.

One morning the scolding stops. Dreading the reason, I look out the cabin window. A pile of blue feathers colors the grass below the infamous limb. Quixote's number is up.

A friend, recently divorced and needing cuddles, happily takes Quixote in. She has a good-sized house, a yard full of ubiquitous starlings, and an advanced degree in cat psychology. She feeds Quixote fresh fish and milk daily, buys toys for him, and lets him sleep under the covers with her. Clearly, a match made in heaven. When I visit, he looks fat and placid, and turns his head aside.

I miss my intelligent companion who could jump on roofs and open doors, his luminous eyes watching me from the foot of the bed, and even his magnanimous way of shuffling all the neighbor cats in to feast at his food dish. Quixote earned his name as a kitten, from his noble scuffs with my socks and dinner napkins. Everything became a gallant battle, and I, even at that low point in my life, his Dulcinea.

But the finches and nuthatches are back in the meadow. The birdbath brims with feathered flips and flutterings. Seed scatters from the crowded feeder. The wildlife has approved of the move.

Sometimes I wonder about the brave jay who brought attention to the cause. Maybe there are martyrs in the bird world?

I keep a blue feather above the door, stroke it for help with hard decisions.

Resourcefulness

The bathroom door swings loose so I braid cordage from rushes to create a latch. This takes all morning and gives my fingers a rash, but hey, it adds character to our home, and I learned something.

Being a gazillion miles from anywhere (okay, twenty miles from the nearest town, but I'm dedicated to driving as little as possible) makes me clever with the resources available (be they old shoes, tree sap, bike tires, river rock). Wild mint leaves make tea, flowers adorn salads, syrup replaces sugar. I rig up a ladder with an old log, a fence from strewn branches; chairs hold up a clothesline, a broom arises from grass.

Two flat stones bookend my library, which rests on a plank I found in the barn. Clothes hang from a twisted branch nailed across the back wall. A basket adorns the front door, made from the pliant, but invasive, English ivy, with a teasel pet comb inside. Teasel was brought west as a tool for teasing wool before spinning it.

Gradually labels break down—socks transcend their lowly status and make startling headbands, scarves become halter tops, window screens turn into drying racks, and, of course, for large parties, a dryer lid makes an excellent pie pan.

Categories dissolve! Anything is possible! I begin to look at objects for their form and material instead of their culturally prescribed use. Using imagination to break through preconceived ideas frees me of the tight jacket of cultural convention and makes me a creator in my world. Of course, people may suspect that I've finally gone over the edge ("she's wearing that rice bag on her head again!"). You have to be comfortable being different, it seems, or have your own community of wacky, resourceful friends.

Although much of modern life is complicated and requires learning intricate details such as how to retrieve all the important data you just lost on your computer, mostly there's a method, a formula, that someone else has created that you need to decipher and devotedly follow. This is a different mental function then searching your property for something to hold your pants up, and far less entertaining.

And it's great retraining for an era of dwindling resources. Reduce, reuse, and recycle make the difference between drying those delicious apples for winter or not. When you live with little means, each of them means a lot.

Trespass

There is a pond in a nearby meadow. When I swim there, small bright green frogs line the shore. I swim toward them, eye to eye, and watch their cheeks quivering steadily with breath until, at the magical closeness, they simultaneously dart into the water in scattered directions, some over my head, splashing in around me. It seems I am somehow baptized, or cursed, by frogs.

I've never seen other humans at the pond, although a worn rope swing hangs off an oak, and a moldy kickboard rocks at the shoreline. Sounds travel easily over the surface, the "kerplunk" of a turtle that lifts its beak, senses me, and slides off its log perch into the water. I float belly up and listen as marsh wrens scold from the reeds: one is low and steady; another pitched, frantic; still another creaks. Clearly, animals have unique voices, recognizable to each other. This was a conversation among individuals, undoubtedly, about me. On a low branch stands an indignant jay, blue head feathers puffed up. It adjusts its position on the branch and stares.

So this is what it is to trespass. To enter another's world uninvited, crush the plants, rip through the unseen spider web, wake the turtle, scatter the frogs. To go where we are not necessarily welcome, where we have asked no permission, oblivious that other beings have a sense of home, which we violate until we either enter respectfully, or make it our own.

Not that we are always trespassers in nature. People have lived as part of the land, moving in without rearranging too much furniture, listening and tending and giving way. For most of us, though, rather than wake in it, we "go into" nature, living as we do mostly indoors.

I will return to the pond daily, all summer to swim, following the same worn path and drying off on the same hot rock.

Eventually, I learn to enter unobtrusively. Watch my step. Go slow. The frogs become laissez-faire, the marsh wrens amp down to a rustle, turtles raise their crusty heads, but stay on their rocks.

Siesta Lane

So goes this crazy quest—for a Chicago gal to put her feet in a pond in Rainbow Valley, Oregon, as a member of the family.

Second Glance

Although "zu-weet" rings at you
from everywhere in the forest,
it takes awhile to find
the pale-green Hutton's vireo singing
from the oak tops. Each trill
causes its whole body
to vibrate. After that,
you cannot stop seeing it
quaking in the golden leaves
every time you look up.
And even with careful steps, you almost miss
the western lizard lumbering across your path
through the elegant jeffrey pine.
But when you bend down,
how startling it stands out
from the needle mulch.
Lifting its curious eye,
fingers splayed on a twig,
clutching, and blinking.

At first you think Oregon ash bark
is white as alkali until
you touch the pale crustose lichen
stretched like a glove
photosynthesizing and reproducing
along its entire trunk.
So when someone proclaims
a woods devoid of life

Amy Minato

egg hatch, grub munch,
bee buzz, frog hop
newt mate, spore slide,
root suck, leaf drop.

it is then you begin to understand
what within us is ill
and wants culling.

Rot

Today I spread the gluey, gooey ripe compost out across the garden and dig it into the soil before planting seeds. The land swallows this dark muck like the good medicine it is, with a squishing sound like a burp.

All winter we have scraped broccoli stems and orange peels off our plates into our compost pail in the kitchen along with coffee grounds, egg shells, and any organic matter except meat, watermelon rinds, and corncobs, and trudged it out to a big open bin near the garden to "feed the critters." When it's pouring rain and the pail slops against clean pants this is not a favorite chore. A cascade of colorful, odoriferous food scraps tumbles onto the grubs and fruit flies that multiply in an orgy of leftovers. I then toss a handful of grass clippings over them the way a priest sprinkles holy water and head back to the house, pail swinging.

Somehow this stinky, writhing pile will convert by spring into a nutrient-rich, brownie dough–like miracle food for plants at the same time that it keeps our garbage to a minimum and trips to the dump rare.

Jazzed by the transformation, I gaze down into the compost bin for clues to the meaning of life and develop a compost philosophy. Since dead bodies, if left alone, will decompose and become part of the life cycle again, maybe souls do the same. Maybe when I die my spirit will break up into little parts and get transformed and rear-ranged and somehow become part of future souls? This could explain past-life memories, spiritual resonances, and déjà vu.

When I explain this theory to Raul, he offers all his future compost duties to me as a gesture, he explains, "toward my enlightenment."

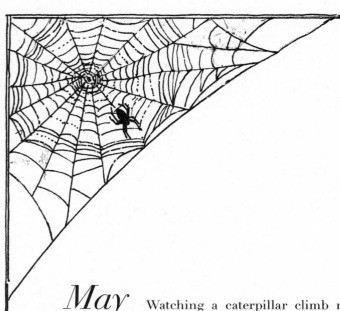

May Watching a caterpillar climb my scarf I marvel at the transformation it will make, wondering at how much we fear our changes, rather than enjoying the magic. The caterpillar travels randomly, dropping down on the first available place, seemingly without attention to pattern, texture, or direction, waving its head around first, then landing, decidedly, arch up.

Reverence

So every dormant bud, water skipper, screeching jay, is an angel, buffeting up beside me. Humble thought. How can I survive the machine age if attending to the voice of every beetle or cherry tree? Let alone one *branch* of the cherry tree, or one chip of its bark, teeming with life.

I embrace the web sustaining me. The sky orients, seas fill and surround, oxygen circles through vegetation, soil collects the least flakes of my skin and hair. Where did I think I was, anyway?

If not already home.

On such days I imagine the Dutch shopkeeper Antoni van Leeuwenhoek in the mid-seventeenth century. He assures Mrs. van Pelten that her fabric will arrive tomorrow, adjusts the shades, locks up, and walks down to have tea with his old friend Carl, the eyeglass maker.

"I suppose, Anton," Carl says, removing his protection mask and looking up from his needlepoint tools, "you'd like to collect some more shards from my floor." He looks amused. "Help yourself."

Leeuwenhoek nods, grins, and sweeps a few crystals into a bandana. "Goodnight, Carl, and thanks." He's up late, grinding glass into finer and finer lenses.

Siesta Lane

He doesn't notice his fatigue, because he's working on a microscope.

Alone in his quiet shop, by lantern glow Leeuwenhoek places a drop of water "just for the heck of it" under the scope, leans over, looks through it, and knocks his socks off.

Paramecium, amoebas, bacteria—"animalcules," he calls them, as they are obviously alive. These tiny creatures in their translucent beauty skitter across the water and he, Antoni van Leeuwenhoek, is the first person ever to see them. He is awed and terrified. Disease will begin to make sense. And water is never quite the same.

What he couldn't have guessed, back then, is that we humans started out similarly. One-celled beings with an attitude. When a prokaryote cell swallowed a mitochondrion four billion years ago, a pact was made. "Carry me around and I'll light your fire, baby." Stored energy. We were on our way.

Leeuwenhoek lived to be ninety-one years old, rarely leaving Holland, enthralled by the new worlds under the lenses: a spider's eye, mouthparts of a louse, capillaries. Why travel?

Sometimes I wonder why we all seem to be looking for discovery only beyond our homes—the Arctic, the ocean, the moon—as if we already know everything there is to know about the dramas unfolding in the compost heap. We forgo the mystery in our own souls and backyards.

That forms of life existed invisible to us must have been a mindblower for Leeuwenhoek. But what today exists beyond our senses? What songs can't we hear? What scents do we miss? What goes untasted by our limited tongues?

The irony is that our ability to detect and appreciate more about the natural world progresses along with our ability to destroy it. The same instruments, sometimes, used for both. What will it take for us to separate these powers? To prove that the apples of knowledge and evil needn't be from the same tree. What leap of consciousness would it take for us to learn to savor one, scorn the other, and loll around in the garden forever?

Appreciation

"Nope. Still green."

We wait in salivating expectation for the strawberries to ripen. Siestans on their way to the mailbox will detour a quarter-mile to check on them. Slugs discovered on the strawberries are, with all due respect, squashed.

When there is the first discernible blush on the berries' bumpy bottoms, we plan menus.

"Let's have pancakes next weekend, then we can have strawberries with 'em."

"We can make strawberry shortcake for Sara's birthday."

"Don't buy fruit! The strawberries are near ready."

The golden carpet of our tastebuds is laid out waiting for those royal red strawberries to parade juicily across.

Finally they are ripe! We gather in the main house with ice cream and peppermint tea, count out twelve (of approximately equal size) per person, and savor them.

Rainbow Valley experiences ten minutes of silence but for sighs, swallows, and metal spoons clinking the bottoms of bowls.

Then Raul wipes his vanilla lips and tells about his mom gathering roots in Idaho. Luke remembers spending all day making tamales at Christmas. Sara, Rita, and I swap jam-making tales. Mick chimes in, jam being half his daily fare. All our domestic stories pull up chairs around us.

We are, it seems, enjoying the enjoying of the berries, even more than the berries. Now, the irony is that we can buy any fruit from mangoes to kiwis at the local store any time of year. We can go to a restaurant that would serve strawberry pie, tarts, or flambeau. But we are instead sitting on the porch at Siesta Lane in the early summer dusk with the new moon arcing over the oaks, jowls crimsoned with juice, rediscovering our capacity for bliss.

The Goat's Eye

The Illinois landscape is to Oregon landscape what a
Jacuzzi is to a hot mineral spring. One strives to imitate a wilder nature, the other still is, at least in a few
places. In Illinois, people work to reclaim their native
landscape. While lifestyles contribute to global destruction—prairies appear in backyards and outside museums. The native landscape has mostly been altered and
naturalists work to bring some back, like a gift. Restoration is the Midwestern environmentalist's main job—as
fixer, healer.

But we have some wilderness here in the West and the
land is still under attack, the wounds fresh. Here nature
lovers are teachers, mourners, or warriors—educating
people about threatened places, writing elegies to beloved
places, standing with placards in front of bulldozers,
squatting in trees, hiring lawyers.

At least we still have a chance to save some of it, and
for those of us who care, this responsibility adds weight
and meaning to our lives. There's an ache to our lives,
sitting here outside the emergency room as we do, nervously flipping through magazines, different than the

Amy Minato

bittersweet reclamation of Midwestern lands. In Oregon, you visit an old growth forest and the next week it may have been logged.

The assaults are constant, daily, and for every tree saved, it seems, a hundred fall. We walk among newly ravaged places like those scenes in the film *Gone With the Wind* of a few doctors wandering among thousands of dying soldiers. In Timothy Egan's book *The Good Rain*, he describes a radio announcer's horror at the devastation below as he flies near the Mount St. Helens eruption site. "The volcano didn't do that, sir," his embarrassed guide whispers. "Those are clear-cuts."

In Illinois, for the sake of diversity, I had let my yard go wild. Neighbors complained until I put up a cute fence around this new meadow and a neat little "Prairie Preserve" sign. The "weeds" were suddenly acceptable as "prairie plants" (like how nasty "swamps" become precious "wetlands") as part of a restoration effort.

In the West we try to ward off destruction with our red and yellow signs—Slow, Stop, Yield. A less glamorous job, this preventative care, but a privilege to witness ecological health. To swim in a glacial lake or yodel back to a wolf, to meet the mountain goat's gaze, and hold it.

The Slough

Wanting to nurture a healthier relationship between people in our community and the land (and enamored with Joseph, the sweet guy with the lush ponytail and olive dark eyes that I shared a couch with at the planning meeting) I help form a group to encourage direct experience and stewardship of the natural world.

Through Nearby Nature we plan to lead kids on nature walks and coordinate volunteer restoration work in local natural areas. We become a family of partial dropouts from the mainstream who share a vision. It involves a weekly trip to town but it's ethical, hopeful work, and it's outside. We turn our first attention to the city's pinched capillary—Amazon Creek.

Amazon Creek, which spindles along through Eugene, was once a veritable river, flexing and sighing in its bed. Since 1959, the floodplain has been filled for development, and the creek channeled. But rivers need to breathe, to replenish and be replenished by nutrients from floodplains. Small ponds form when the river recedes, providing vital sanctuaries for tree frogs and pond turtles. A river allowed to spread to its natural girth will create islands of habitat for heron, beaver, mink.

Amy Minato

Amazon slough, however, is now an algal sludge trickling along the bottom of a deep scar, dredged by the Army Corps of Engineers. Throughout the summer, grasses and rushes are relentlessly mowed to the very lips of the creek, destroying habitat and food. The trees in the Amazon park are "bottomed" to give shade and to keep transients from living in the skirts. The trees look naked and imbalanced, shamed in a way. All done in service to a standard of beauty that can be traced back to America's British roots: a preference for manicured, ordered gardens and a fear of insects. In the Anglo tradition, managing, controlling, clipping, mowing, or chopping gives an area value. It bears our mark. It has yielded, become safe.

But to survive we must learn to loosen up, mingle with bugs, revere the tangle of underbrush, the individual shapes of trees and tall, teasing grass.

I bring a group of Nearby Nature children to the park. Fortunately, there is an unmowed area beside an island of trees. First, we pay homage to the robin puffed and squatting in its nest in the ecotone between the woods and meadow, dark eye gleaming.

We play, hide, and laugh in the swath of grass taller than the kindergarteners. We caress a caterpillar, whistle across dandelion seed, tickle each other with fescue. Mauve pollen colors our hair and clothes. This small meadow becomes our secret place, shared with life forms too myriad to imagine.

While we play, the park mower roars in the distance. It frightens us, more so with our new awareness of the fragile ecosystem webbed in the tall grass. Together we imagine how it compacts the soil, clips the plants before they can make seeds, chews up snakes. The kids each collect a favorite plant stalk to press and keep. We wave back at the dancing bromes as we leave.

Siesta Lane

I return the next week with another group of kids and prepare to send them to a spot hidden in the grass, to treasure-hunt for life. But the area where we play has been mowed. It is now as bald and yellowed as the rest of the park, but for a patch the size of a pool table where the Nature Conservancy has found a stalk of endangered tufted hair grass. This swath stands out like an exhibit, a museum piece to remind us of something lost and gone. Eight swallows sweep over this vestige of insect habitat. The robin has abandoned her nest, probably to seek out a better food source.

There is no one around to confront, just the parking lot and impervious picnic tables. I am stunned and silenced. Had I not played in this meadow last week, my response to the mowing would have been mute, if I'd even noticed it. Not having rolled around and napped here, the news might have brought me only mild irritation, instead of a gnawing sense of loss. I call park

officials but they get more calls from folks who want the park manicured, tidy, "under control."

For change to happen, maybe more of us need to get out of our cars, off our lawnmowers, and down onto the dirt. Making connection with a place brings familiarity, responsibility, and an extension of self. There is risk in caring about what is not widely valued, but greater risk lies in not honoring what is precious.

Amy Minato

The kids' excitement, buoyed by stories of crab spiders, robin chicks, and camouflage games in the tall grass, quickly deflates. They sit on the ground and cover their legs with the dead grass, contemplating, maybe, what scant world might be left to them.

Reckoning

I am having a tea party with three of my worst faults.

We spread a blanket under an oak and nervously gather around it. Selfish grabs the spot with the best view, Fearful hovers behind a tree, and Careless spills her tea. The fragrant stain spreads across the fabric.

Why did I agree to this, I wonder? Haven't I suffered enough from these characters' surreptitious appearances throughout my life? Upon its arrival I tore up the invitation to this sorry event. I threw it in the fire, ignored it on the shelf. But nothing worked. There would always be a new one delivered to my door. Crisp and foreboding.

So I brew tea and bake cookies. We meet outside surrounded by the supportive presence of grass and squirrels.

"Selfish is hogging the blanket!"

"Careless stepped on my foot!"

Fearful shifts her gaze down and passes out napkins.

I peruse these pathetic aspects of myself and sigh into my tea. How did we get here? I'd meant to eradicate these flaws by now. But look at them. They need me.

Amy Minato

Instead of rejecting them I look deeper. At the core of Selfishness stands a guardian, in Fearful resides a peace-maker, in Careless a dreamer.

They sit patiently around the blanket, Careless with crumbs on her shirt.

They will not go away, nor should they.

From behind an oak, a willowy figure approaches. Dressed in overalls, hair tousled, Acceptance sits beside me. She holds my cold hand. We begin a conversation that may take my life to finish

Respect

Through Nearby Nature I begin a group called the Green Scouts, a children's club that does environmental restoration activities. One day, as we collect rare, native mule ear seeds for the Nature Conservancy, a child finds a spider. A large, colorful one in an enormous web. Kyle has an immediate, fervent desire to own this spider.

Where are the stories, I wonder, to help us learn a simple respect for other life? To teach us to take with gratitude, and only out of necessity. Where are our rituals? Native cultures are rife with them, woven into the rhythm of their lives. Stories that warn about greed and hubris, rituals to mollify the horrors of the hunt. Today many of us move from instinct and buried culture, borrowing from others the social tools for creating a spiritual and sustainable bond with the Earth.

So I explain to Kyle how spiders are important in the food chain, how they eat and are eaten by other species, how much better it would be for the spider to stay right there at Willow Creek. We talk about how we do these activities to help nature, how for one morning a month, we put the needs of the plants and animals above our own. He whines and resists until I give up, and he carries the spider toward his mom who has come to pick him up.

But just as he reaches her he says, "Wait a minute. I gotta go put this spider back. It'll be happier here." Kyle's manner when he lays the spider back on the grass is different than his glee when he'd found it. He puts it back gently, with a serious look, more than a little proud of himself. Kyle watches the spider scurry off over dandelion heads, and he looks pleased at himself, his seed collecting a small feat in comparison.

June I sit on the porch at dusk with the moon between the oak, throwing shadows over the long meadow and song just comes, low and easy, out of my throat to mingle with cricket chirp and owl moan and the staccato of nearby dogs. Low rustlings and furtive twigs cracking and suddenly the woods are dotted with glowing eyes. So still, settled into myself for so long. Here, like a sentinel, watching things happen.

Song

One forest ecology theory postulates that as the motion of their sap slows, diseased trees in the forest vibrate at a faster rate, producing a higher pitch than healthy trees. Insects harken to this particular tone, congregating around the sick tree, a phenomenon that serves to prune the forest.

Now when a bird calls, I'll abandon my current task to find it. I'll leave the door banging, tune my ears to direction and height. For low-toned songs I'll look down, for shrill songs, up. A hermit thrush hanging out close to the ground sings in lower tones than high-pitched treetop songbirds like the winter wren. Low sounds travel farther through dense woods; high tones vibrate easily above trees.

Siesta Lane

I'll separate the sound and try to imitate it, to respond, to seek out the bird in the branch. Remembering that one sounds like a stutter, another a whistle, another a scold. Meanwhile my toast is burning, fire sputtering, telephone left ringing in the house.

Maybe this is a small knowledge of what it means to bring the background to the foreground—to honor the natural over the human-made, to know where our temple is. And so maybe if we stop and listen, life will tune and strum more insistently, calling itself awake. Frog chirp. Cricket call. Fire crackle. Bee hum.

At dusk last evening on a telephone wire, a starling was practicing a warbler's song. Starlings, which are related to mockingbirds, mimic the calls of other birds. No one knows why.

Starlings were brought from England because someone decided that the New World should host all the birds found in Shakespeare. Today across the country, they continue to crowd out many native birds. Starlings take over nest sites and food sources, intimidating smaller, weaker species.

This one would start over and over, each time making small changes in the sound. When it finally settled on a tune, it repeated it many times, not quite capturing the warbler's easy lilt. Another starling stood below, looking up, seeming to listen. I imagined that starlings mimic for the purpose of tricking their way to a nest site, or to fool an amateur naturalist like myself. But maybe it's from a nostalgia for what in nature they are eliminating in order to thrive. Maybe this gesture is a requiem, as are our films, records, museums: an attempt to secure a song that, just by living in our particular manner, we are extinguishing.

Niche

Often during my work with kids in Amazon Park I watch the drama of the red-winged blackbird enacted against the backdrop of joggers, cyclists, and Frisbee players. While a common bird in many places, there are few red-wings in the city of Eugene and only one pair of them left along this stretch of Amazon Creek. Today, one is chasing crows away from its nest in the willows and cattails. A bird as large as an oak leaf chasing one as large as a kite. Three more crows speckle the mowed lawn beside the creek, waiting. They are patient and implacable, their slick feathers cobalt blue. The two blackbirds take turns chasing and nest-sitting, and exhaust themselves with their valiant pursuit and scolding, the red on their shoulders flashing like wounds.

It seems impossible that they have time to hunt, and that the chicks are getting enough to eat. I don't believe we will see them here next spring.

A woman who grew up near this park forty years ago says turtles were abundant in the creek, which used to spread ten times its present width. Each morning she'd wake to the song of meadowlarks thick as mosquitoes. But the Amazon today is pinched into a single vein and the ponds where the western pond turtle laid its eggs have disappeared. The turtle eggs that manage to hatch today are eaten by the non-native bullfrogs.

Without the tall grass we lose the meadowlarks and their liquid songs. What other life forms no longer grace Amazon Creek? What else is lost that might wake us to beauty, quicken our senses, stir our imagination, and woo us toward love?

Among species there are generalists and specialists. Generalist species introduced into a disturbed ecosystem tend to have advantages over native species. Maybe they are adaptable to disturbed conditions, or immune to local diseases, or more aggressive. Generalists are competitive, tenacious, pervasive. They can adapt to a variety of environments, climates, and foods. They are cockroaches, crows, bullfrogs, possums, blackberries. They are the bane of the specialists.

Specialists need particular environments—clean air and water, shelter, certain foods. They tend to be more passive, sensitive, and yielding than generalists, and are winnowed out first when an environment changes. Unable to adapt to newer, usually harsher conditions, or to compete with colonizing species, specialists retreat, decline, are quietly extirpated.

As one who requires a natural habitat, I root for the red-wings. We human "specialists" also try to ward off destruction and sound the alarm—sensing what in us can survive only in a living environment, preferring mountains to malls, streams to subways, air to airplanes. Can we read our own future in the story of the red-wing blackbird? Are we putting up as gallant a defense?

Succession

A few minutes after telling a visiting boy that the tree house is off limits, we hear an ominous crash. He has fallen from it and broken his arm.

The distraught parents threaten to sue the owner, whose nervous response is to pick up the pace on selling the property.

This, of course, freaks us out. I even consider buying the land, which prompts a disgusted grimace from Raul, who doesn't believe in owning land.

Still I concoct elaborate schemes and calculations to warrant doing it. If I never so much as have to fix a leaky pipe, maybe I could squeak by on my pathetic savings.

But the place needs work. With eight structures to care for, it's a maintenance drain. And for someone as unhandy as myself, a likely nightmare.

Yet I've never felt so at home anywhere, so right about a place. I need to watch lichen wrap the oaks, barn swallows teeter on wooden fences, clouds dab the far hills.

If I had someone to buy it with, maybe, or a few others. Mick, Sara, and I kick around the idea of shared ownership. The logistics overwhelm us and it no longer seems like a simple life. The well may be drying up, the septic field is nearly full, buildings must be brought to code. What if one of us wants out?

And here's the catch: Living in the country is one thing. Owning country property quite another.

Gradually at house meetings I notice a disharmony among us. Some begin to see others as possible landlords and the camaraderie fizzles. Luke chafes at chores and Rita whines about the phone. Usually, house meetings are joke and dessert fests. Now power has laid his big stick on the table and the community spirit I love catches her gown on the windowsill as she scurries out.

We just want to continue as is, serene citizens of Oz with our Wizard in the next room. But someone will buy the land and we'll all have to move. The meek may inherit the earth but only after the rich have died off.

Douglas fir trees rim the oak meadow: implacable, patient. They know that according to the laws of succession they'll move in on grasses and shrubs. Conifers grow taller and shade out deciduous trees. They send their stiff, pointy soldiers into the field each spring.

What keeps them at bay are fires, or mowers. Established oaks with their armor of bark can tolerate fire much better than Douglas firs. Still, each year the border of conifers advances into the meadow. It's only a matter of time.

Californians looking for a quaint retreat home come to check out the property. Bewildered by the number of cabins and by the number and look of us maybe, they decide not to buy it.

In the dim light, the dark line of conifers appears to recede from the house. We temporarily return to ignorant bliss.

Courtship

Meanwhile, my mind and other parts circulate around the idea of Joseph, with his bicyclist's body and charcoal eyes that smolder at me over his glasses as if waiting for a reply.

For the past few months everywhere I look things are mating. Dragonflies hook together over the pond, swallows tilt in pairs through the sky, squirrels careen over each other along tree trunks.

Garter snakes go into brumation, where they stop eating for two weeks to clear their stomachs before sex. Pheromones slinking off a female's skin attract male garter snakes that roll around her in balls of up to a hundred snakes.

Even the plants seem aroused. Cattail stalks wave their corn dog tops while a chubby bee with a gold-dusted abdomen snuggles into a smug iris. They are my cheerleaders, cueing me toward love. I clean out my cabin, skip meals, and caress Joseph's phone number taped to a photo of an otter above my pillow.

Amy Minato

Aerial

We could begin
as dragonflies in fringed hover
lace wings in rapid spiral
abdomens throbbing
the myriad facets
of our large eyes aglow.

Or as swifts
cupped by wind
caressed by cloud
feathers smoothed to silk
in the sky's palm coupling
and uncoupling.

We could fold flesh
curved and furred as bats
fast and warm-blooded waving
and snapping like castanets.

But finally I would want
to fly in fierce orbit with you
surrender to gravity clasp talons
scream through hooked beaks
and hurtle toward earth falling
like eagles in love.

Nomads

The Siesta Lane property has been sold to a woman who
lived here before, who moved to town and came limping
back. She's working several jobs in town to afford her
daily minutes of peace here. She's also raising the rent,
and cleaning house. We have to move out by June.

We hold a house meeting to discuss our options. I'm
the only one without a plan, the only one who has sol-
dered myself to the place—Scarlet O'Hara clenching her
clod of dirt.

Sara will go to Minnesota, Jack and Rita to England;
Luke will take a studio in town, Raul a job in New Mexico;
Paul and Natalie want to live nearer to her school; Mick
will marry and move to a house in town. At our last meeting, Raul plays his drum and Mick brings flowers. Sara
pours sweet tea into tiny cups. Natalie crochets goodbye
washcloths for everyone. Luke looks around with the soft
sympathy of the very shy. We exchange well-wishes and
addresses.

It's them I'll miss as well, it's "us," our community.
The squirrels and grasses and oaks and people who share
the riches and secrets of this land.

Amy Minato

Nerve endings flare—my alliance to this place rages like a rash. I lie awake at night, heavy in my bed, under the moon and branches, feeling the pull of each and every vein growing from my body into this land. Imagining that the weight of my desire to stay can somehow root me to it.

But the owner paid for the place, so it's hers. She has power to affect what happens here, which bonds are formed and which severed. She mowed the meadow right off, and yesterday we found half a bunny head on the living room carpet, a gift from her cat.

Leavetaking

It is our last day at Siesta Lane, this place of rest and replenishment in the middle of my life. Joseph, who will soon become my lover and eventually my husband, sits next to me on the porch swing. He is providing me moral support and chocolate chip cookies.

Joseph, whose unenviable task it is to teach me all the physical science I should have learned in high school, tries to distract me with an explanation of geologic time. Tertiary, Quaternary, Miocene. These sound, to me, like heart muscles, which reminds me of strains and strokes, which brings up pain and loss.

I don't want to go.

"Look at it this way," he says. "Siesta Lane has taught you how to be in a place, how to really love the land. That capacity is inside you. You'll always remember this place for that gift, which you'll bring everywhere you go."

I squint and try turning the prism of my sharp hurt to see if it will flash such rainbows. Maybe. Maybe.

We watch everyone else briskly and methodically carry boxes to their cars.

"Bunch o' bees," Joseph says, gently kicking the swing.

Oh, yes! We are like bees leaving in a swarm for new lives.

And what honey will we each make from our time here, what flowers will be fertilized with the pollen stuck to our fortunate, bewildered souls?

"Will you help me with my bed?"

"Of course."

Beyond Siesta Lane

Hearing about Thoreau, one gets the impression that he lived the whole of his life at Walden Pond. Actually, he moved back to town after two years of his experiment. I've often wondered how he reconciled such a change.

Years later I live in a tiny house behind a large field, in Eugene. A long-awaited child grows within me. Joseph and I work together to encourage appreciation and stewardship of nature and the practice of simple living.

We try not to use our single small car. Nearly everything we own is secondhand and our diet relies heavily on organic, locally produced food. When possible, we buy in bulk and use recycled paper. We are lucky to live in a place conducive to our ways.

Still, we use phones and computers, refrigerators and washing machines, even as our hydroelectric dams squeeze out salmon populations and industry waste poisons our land. Homemaking has invited more belongings into our lives. It's an imperfect balance, but we walk the tightrope with our hearts in our throats.

Amy Minato

We attempt, at least, to move in the direction of walking lightly on the Earth. Because we believe that the key to a profound life is lower expectations of quantity, higher of quality, we strive to have few possessions that we take good care of, and to do less, better.

How can I walk to work and stop to watch buffle-head ducks skim a pond if I have ninety more errands to do that day? Fast food culture and the automobile depend on the desirability of convenience, and convenience is mainly important when you're trying to do too much.

Slowing down my lifestyle was easier in the country, where time passage is measured in cloud drift and shadow shift. In the city, it's hard not to collapse exhausted in front of the TV every night, but finding ways to cut down on stress reduces the desire to zone out.

In our culture we're taught that more is better, and conduct our lives accordingly. A Colombian friend once remarked, "When I ask Americans, 'How was your day?,' they reply either, 'It was a good day. I got a lot done,' or, 'Lousy. I hardly got anything done.' In my country, our answers would be more like 'I enjoyed today. I walked with my sister,' or, 'I'm sad because my friend moved away.'"

As Joseph and I lessen our "to do" list, we're plagued by guilt and a sense of inadequacy—puritan residue. Until we remind ourselves that overproduction is what's hammering our planet.

I buy a small daily planner that will fit only a few obligations per day. We figure a small house is quicker to clean, a yard turned to wildlife habitat requires less work. The time freed up allows us to volunteer for good causes, cook food from scratch, give massages, watch birds. Buying only essentials (and these secondhand) makes high-power, high-stress jobs less necessary.

Siesta Lane

Sometimes, as we bike along in the cold rain, passing drivers give us pitying looks. But we wouldn't switch places. Even if we were tempted, the child within me, rocking closer to life with each turn of the pedal, would have none of it.

Acknowledgments

Thanks to those who read Siesta Lane and offered tactful and sensitive feedback even when it called for major surgery: Diana Abu-Jaber, Ed Alverson, Mary B. Fort, Alan Dickman, Cath Fleischman, Jenny Koll, Donald Snow, Barbara Scot, Jack Shoemaker, Stephen Williams, Mary Wood. Thanks to my agent Janet Reid, for her faith in the book, editor Ann Treistman for her sensitive suggestions, copyeditor Jesse Shiers (sorry for all those extra periods!), and Jan Muir for her inspiring illustrations.

Thanks to David Memmott and Ice River Press for extending rights to reprint the poems.

Thanks to Beth Stein for taking me to re-visit Siesta Lane and for sharing her exquisite photos of the place.

Thanks to all the courageous, creative folk dedicated to leaving a healthy home for our descendants, alone or in groups such as Aprovecho, Nearby Nature, the Center for Appropriate Transport, Community Supported Agriculture, The Nature Conservancy and others.

Amy Minato

Thanks to my parents, siblings and other members of my large, extended family and network of friends who are my sustaining web.

And in appreciation of folks who have ever lived at Siesta Lane, and those who aspire to—in flesh or spirit.